Nonhuman Humanitarians

Nonhuman Humanitarians

ANIMAL INTERVENTIONS IN GLOBAL POLITICS

Benjamin Meiches

UNIVERSITY OF MINNESOTA PRESS
Minneapolis
London

A portion of chapter 1 was published in a different form in "Nonhuman Humanitar-ians," *Review of International Studies* 45, no. 1 (2019): 1–19.

Published by the University of Minnesota Press
111 Third Avenue South, Suite 290
Minneapolis, MN 55401-2520
http://www.upress.umn.edu

ISBN 978-1-5179-1384-7 (hc)
ISBN 978-1-5179-1385-4 (pb)

A Cataloging-in-Publication record for this book is available from the Library of Congress.

To Bebhinne, Adira, and Emmett, for
love that is mighty and true

To Mollie, for encouraging me to think
about justice and animals

Contents

Introduction

THE HUMANITY AND INHUMANITY
OF HUMANITARIANISM

And then, about halfway through our long captivity, for a few short weeks, before the sentinels chased him away, a wandering dog entered our lives. One day he came to meet this rabble as we returned under guard from work. He survived in some wild patch in the region of the camp. But we called him Bobby, an exotic name, as one does with a cherished dog. He would appear at morning assembly and was waiting for us as we returned, jumping up and down and barking in delight. For him, there was no doubt that we were men. Perhaps the dog that recognized Ulysses beneath his disgust on his return from the Odyssey was a forebear of our own. But no, no! There, they were in Ithaca and the Fatherland. Here, we were nowhere. This dog was the last Kantian in Nazi Germany, without the brain needed to universalize maxims and drives. He was a descendant of the dogs of Egypt. And his friendly growling, his animal faith, was born from the silence of his forefathers on the banks of the Nile.[1]

The work of Emmanuel Levinas, renowned philosopher and Holocaust survivor, is an unusual starting point for a discussion of humanitarianism. Yet, Levinas's work on ethics, in particular his analysis of one's responsibility to the other, offers perhaps one of the most radical, uncompromising arguments in favor of humanitarian aspirations.[2] For Levinas, thought, compassion, empathy, reason, identity, and care flow from the overwhelming presence of the other, an other that precedes and constitutes the condition of possibility for an individual's existence.[3] Crudely summarized, ignoring the plight of others, focusing on the self, inquiring narcissistically into the meaning of being, or seeking solely to better one's own life amounts to an erasure of this singular responsibility. Such an erasure is not just unethical; it is a betrayal of a relationship so fundamental that it underlies human facticity or the mere fact of existence.[4] As such, Levinas contends that attentiveness to the other is an imperative that no political, religious, economic, legal, personal, or social obstacle should impede. In making this claim, Levinas hypothetically provides one of the strongest justifications for humanitarianism because he demonstrates that any argument for ignoring the suffering of the other based on nationalist, ideological, economic, or political rationales is little more than an unethical alibi.

Despite his articulation of this ethical priority, Levinas rarely appears in popular discussions of humanitarianism. Rather, humanitarianism is typically presented as defensible from one of three different perspectives. First is a set of theological or moral standards that emphasize the importance of charitable action.[5] Second is a set of observations about the illegitimacy of pain. In particular, with the rise of the Enlightenment, modern humans allegedly have the capacity to use reason to determine and denounce the irrationality of suffering.[6] Here reason produces a new condition that can expose the arbitrariness of misery and the fallacies that justify cruelty. Finally, humanitarianism is often understood as an outgrowth of a natural human capacity for empathy.[7] In this explanation, humanitarianism results from a collective, historical process of emotional growth that signifies the possibility of a future condition of peacefulness. In this case, it is this ability of humans to share similar sentiments or feelings, a presumptive global community united by affect or emotion, that explains the importance of humanitarianism as a strategy for alleviating injustice.[8]

Each defense emphasizes distinct attributes of humanity and rationales (theological aspiration, rational explanation, or empathic nature) to explain humanitarian responses to the suffering of others. Yet, these claims emerge within a similar historical context, share a vision of ethical responsibilities transcending other duties, and adopt resonant visions of global politics. For instance, religious justifications for humanitarianism materialize from theological debates and presuppose a model of universal suffering that resembles the image of grief found in sentimental accounts of humanitarianism. Progressive denunciations of inhumane treatment assume that humans can identify and criticize illegitimate forms of pain because of a common affinity for thought that links the capacities for rationality to sensation and feeling. Put differently, underlying these three justifications for humanitarian action is a paradigm that understands human life in universal terms, highlights the inevitability of human suffering, underscores the importance of emotion as a conduit for politics, and defines the human condition as ethical because of a capacity for mutual recognition and care. These are strong, if contestable, explanations of the origin and value of humanitarianism.

Levinas, however, offers a potentially more persuasive argument. By situating the duty to care for others beyond any question of shared humanity, capacity for reason, or emotional affinities, Levinas makes the problem originary. For Levinas, the other precedes the self ontologically; it makes an "I" and an "us" possible and, correspondingly, constitutes the sole horizon for ethical responses.[9] Traditional religious, rational, or emotive justifications for humanitarianism, in contrast, rely on a double gesture: affirming the precarity or suffering of the other while also viewing that other, in some fashion, as fundamentally similar in religious, rational, or emotional terms. Inspired by these beliefs, humanitarian action becomes more inclusive or exclusive depending on how open the advocate is about their underlying notion of human life. Religious or political creeds may thus support strong models of humanitarianism that aspire to universalism or cruel, exclusionary, colonial doctrines. Indeed, the category of universal humanity, bonded through shared capacity, constitutes both a basis for inclusion and a mechanism of indictment and liability.[10] If traditional explanations account for the rise of humanitarianism by appealing to common human capacities based on a shared affinity for ethical investment in the pain of the other, Levinas transforms this commitment from a common capacity of humans into an existential imperative, constitutive of all human life, in which it is the otherness of the other that necessitates ethical action and not, per se, a common feature or shared attitude. In making this shift, Levinas critiques the implicit limitations that inevitability form in discussions about what human lives matter, why they matter, and what we should do to affirm them. If humanitarian principles are traditionally viewed as a product of a religious, moral, emotive, or rational universe, Levinas reverses the point and argues in favor of something like humanitarianism even in the absence of these sources of commonality.

And yet there is a catch. In the foregoing passage, Levinas openly, if gently, mocks Bobby the dog. His mockery is particularly insensitive, as the late philosopher Jacques Derrida poignantly observed.[11] Levinas derides Bobby as lacking the brain to universalize his drives, ridicules the dog's inability to understand maxims, and credits Bobby only insofar as he acts on a base, animal faith. At the same time, in the center of desolation, death, and indignity, in the place that Primo Levi famously called the "gray zone," Bobby joyfully interacts with people reduced to nothing.[12] Levinas's vignette articulates that joy. He comments that "at the supreme hour of his institution, *with neither ethics nor logos,* the dog will attest to the dignity of the person. This is what the friend of man means. *There is a transcendence in the animal!* And the clear verse which we began is given a new meaning. It *reminds* us of the debt that is always open."[13] In this statement, Levinas denies Bobby ethics or logic. He interprets Bobby's kindnesses as a meager *imitation* of our relationship to the other, a pale *mimicry* of our responsibility to the other, a *reminder* of this relationship, but not the real thing. It is as if Levinas is upset with this dog, as if the confirmation that Bobby *could care* about human others as others debases the recipients of his care, stripping from human beings their one final, allegedly human feature: their capacity to care as articulated for one another based on common human principles. Levinas retaliates in a cutting commentary to which Bobby cannot directly respond.

For the moment, assume that Levinas supplies the strongest, most uncompromising argument for humanitarian principles and, moreover, in this anecdote, that Bobby enacts a type of ethical response to the other in horrific conditions. Why, then, does Levinas persist in his attack on Bobby? What is at stake in depicting this dog as lacking brains, as impulsive, as stupid, as *behaving* according to natural predisposition and animal instinct, rather than *acting* from cultivated, ethical consideration? Derrida again provides an apt reading of this move. For Levinas, there is a "value given to fraternity . . . [but] a face that is first of all that of my brother and my neighbor (however distant or foreign he be). . . . It is a matter of putting the animal outside of the ethical circuit."[14] This is an aggressive aspect of Levinas's thought—a desire to mold the other, no matter how remote or different, in a familiar human form. Ethical responsibility to the other dissipates at the threshold of nonhuman life.[15] For Levinas, Bobby is not challenging the gray zone or contesting Nazi violence; he is simply frothing at the mouth and eager to please. In making this claim, Levinas echoes a strong tradition of dismissing the nonhuman animals as self-aware, ethical agents. Immanuel Kant, for instance, argued that "the fact that the human being can have the representation 'I' raises him infinitely above all the other beings on earth. Because of this he is a *person,* and by virtue of the unity of consciousness through all changes that happen to him, one and the same person—i.e., through rank and dignity an entirely different being from *things,* such as irrational animals, with which one can do as one likes."[16] These types of formulations minimize or eliminate the possibility of nonhuman animal awareness, rationality, and morality and ensure that they "cannot become the subject of rights or duties."[17] Even if Levinas does not follow Kant in suggesting that one can act however one wishes with an animal, his theory, which maintains that ethical responses to the other are the ground of being, reaches a limit in the case of a dog.

What does it suggest if the most radical defense of humanitarian ethics also expresses hostility and resentment at the possibility that a nonhuman animal could enact humanitarian practices or perform humanitarian duties? What is so threatening about the possibility that nonhuman animals practice humanitarianism, that they create ethical relations to others characterized by care, generosity, or joy? To be clear, nonhuman animals should not be lumped into one general category because, as any biologist will tell you, organisms are distinctive, and their interrelationships and interactions constitute a core problem of ecology.[18] The animal, as Derrida argues, is nothing more (or less) than "a word that men have given themselves the right to give . . . in order to corral a large number of living beings within a single concept: 'the Animal' they say."[19] This category is a semiotic error designed to capture life much as it also constitutes a strategic point for contesting specific norms of anthropocentric power. Here, however, the term *nonhuman humanitarians* (and nonhuman animals) challenges an implicit framework, a set of conceptual corrals, inherited and bolstered by humanitarian politics. In this sense, Levinas is an important reference point to discuss the relationship between nonhuman animals and humanitarianism because he is otherwise so insistent on dismissing flimsy rationales and arbitrary excuses for ignoring the plight of others, but nonetheless places nonhuman animals outside the realm of ethical action. Certainly Levinas is a peripheral figure relative to mainstream discussions of humanitarianism and far more likely to be explored in seminars on Continental philosophy or Jewish theology than included in dialogues about humanitarian interventions.[20] Nonetheless, his animosity to nonhuman animals reflects a tendency that is implicit and widespread in humanitarian discourse where concern for the other is defined primarily in human terms and where ethics is linked closely to the celebration of humane attributes. Indeed, despite major differences in policy, impartiality, neutrality, and strategy, when framed in relation to nonhuman others, it is hard to miss consistent emphasis on *human* needs as *human* in the principles of almost every major international humanitarian organization. For instance, the Salvation Army makes a religiously inspired effort to "meet *human* needs." The ideal of "humanity," with the goal of "alleviat[ing] human suffering wherever it may be found," informs the work of the International Red Cross. Doctors without Borders likewise centers its practice on "respect for *human beings* and their fundamental *human* rights."[21] Despite the radical character of Levinas's work, his articulation of the human face of humane action is paradigmatically common within humanitarian circles that tacitly ascribe a difference to human ethics, humane conduct, and human suffering.

This focus on the humanity of humanitarianism has two notable implications. First, it frames suffering as important solely when it becomes a human problem. It is human pain, human injustice, that matters. More precisely, it is the *humanness* of this suffering that explains why it matters and to whom it should matter. This emphasis juxtaposes the human with the nonhuman and inhuman, privileging the former term, demoting the middle, and opposing the later.[22] It removes nonhuman animals from dialogues about justice, ethical practice, social agency, and mass suffering. At this moment, with both deliberate and careless destruction of nonhuman animals and innumerable other life-forms at new extremes, when terms like *Anthropocene* and *mass extinction* are no longer jargon, exclusive humanism obscures some of the most horrifying violence that defines contemporary global politics.[23] In some ways, this overt emphasis on the human is strange because the human is itself a shifting, historically contingent concept. In this sense, the human is not just a biological classification but a political term with a distinct historical origin and transformation.[24] As much as the concept of the human features in calls for universal rights and equity, it has also been deployed to dispose of forms of life understood as insufficiently progressed, uncivilized, inferior, or dangerous typically in reference to race, sexuality, ability, culture, gender, ability, or class.[25] More importantly, if the human is a shifting semiotic, then it does not possess stable boundaries in relation to the nonhuman or inhuman. As such, the stress placed on responding to human suffering within humanitarianism circles participates in a problematic disengagement with nonhuman life. This problem does not only concern nonhuman animals but is equally pertinent to how humanitarianism addresses humans, because the fluidity of the concept of the human produces and sustains inequity within and between human and nonhuman communities.

Second, the focus on humanity within humanitarianism encourages a specific image of the agent of humanitarian action. Put simply, humans make humanitarianism what it is, and humanitarianism functions as a discourse that defines the proper forms of conduct for humans. If, as their advocates claim, discourses on humanitarianism shape normative and regulatory ideals about what it means to be human in a global context, then humanitarianism contributes to the definition of both individual and collective subjectivity.[26] Many humanitarian advocates understand the practice as explicitly raising questions about the meaning and distribution of care, the significance of being human, and the possibility of transforming oneself.[27] This tradition often becomes a conversation about what feature of humanity—empathy, reason, sociality—defines the distinctiveness of human dignity. In this way, care becomes a natural reflex or, conversely, a rarefied attribute of some enlightened humans. Setting aside the considerable problem that the category of humanity has historically been used both to support beneficent intervention and to authorize atrocities, it is empirically difficult to document how these capacities of humans emerged or to catalog the conditions that contributed to their birth from within this framework. Indeed, many humans exhibit disregard, disinterest, and overt hostility to one another. Are these not human forms of conduct? Most humanitarians would likely agree and, upon closer scrutiny, argue that humanitarian action involves an aspiration or a normative ideal of cultivating respect and concern for others rather than an intrinsic or preexisting feature of all humans. They may, in turn, describe human hostility and hatred as stupid, uncivilized, base, or animalistic. But this response generates two further problems. On one hand, it shows how anthropocentric perceptions structure humanitarian discourses on conduct and, occasionally, serve as the basis for legitimating hierarchy within human communities by linking some forms of human activity with a lower station of animal life. On the other hand, if humanitarianism is something one cultivates, if one "becomes" a humanitarian, then what disqualifies nonhuman animals, as well as nonhuman things or objects, from "becoming" humanitarian? Although nonhuman animals may not exercise compassion, care, or reason like humans, neither do all nominal humans exercise these capacities in the same way, nor is there agreement on the most basic meaning of these terms. Humanitarians rarely address the question of why human, rather than nonhuman, capacities are so central to their practice, nor do they engage the crucial issue of how the human differs from the nonhuman. As a result, humanitarianism ends up generating discursive loops: citing the natural character of humane capacities but also claiming that these capacities must be cultivated, contending that all humans deserve dignified treatment but making exceptions for those humans who are inhuman monsters. These problems are a lacuna, an unstated, unthought dilemma, folded into the constitution of humanitarian politics. This lacuna develops because humanitarian institutions and discourses explicitly foreground the human as both the source and recipient of ethical conduct without contending with the instability at the center of the concept.

Humanitarians attempt to resolve this problem in several ways. Nonhuman animals are understood, first, as insufficiently human to be the recipients of humanitarian care and, second, as insufficiently humane to provide humanitarian services. Yet, humanitarianism, like all practices, emerges from a heterogeneous social ecology, one in which nonhuman animals and other forms of life are deeply entangled with humanitarian projects.[28] Furthermore, care, a centerpiece of humanitarian ethical aspirations, is a product of the way in which a form of life encounters the world.[29] While a nonhuman animal's mode of accessing reality (and, consequently, care) may not resemble a human version (whatever that might be), this difference does not imply that nonhuman animals are incapable of generosity or political responses to suffering. To the contrary, nonhuman animals also access reality, play, seek to reproduce elements of this reality, craft territories, and express sadness and joy with others in multiple ways. Some of these forms of care intersect, resonate, and strengthen the types of care associated with humanitarianism. Even a casual glance at nonhuman companion species, such as dogs and cats, reveals that some nonhuman animals have a strong tendency to cohabitate, clean wounds, and forge affiliations in ways that are normative in human communities. The fact that these interactions result from evolutionary forces only provides further evidence that humanitarianism is something that, in the extreme, over time may be molded (for better or worse) and, moreover, that may also have played an influential role in generating prosocial human conduct.[30] This propensity places humanitarianism in a curious predicament. On one hand, humanitarianism celebrates and derives its aspirations from the presumptive generosity of caring for the suffering of others. On the other hand, it confronts an ecological context that provides more and more evidence that nonhumans not only exhibit the capacity to care but often excel at the labor of caring for others, depending on the context. Unfortunately, humanitarianism clings to an exclusively human image of agency and, in doing so, misunderstands the conditions of possibility and enactment of its own practice. As such, humanitarianism can either open itself to this realization, redefine terms, scope, and practices and undergo a process of self-transformation, one that potentially involves moving beyond the framework of the human within humanitarianism, or foreclose the possibility of more complex engagement with the suffering and capacities of a nonhuman world, to the detriment of both nonhumans and humans.

ENTER NONHUMAN HUMANITARIANS

What makes this challenge particularly urgent is the sudden appearance of nonhuman animals as laborers within humanitarian endeavors. Two centuries ago, humanitarianism primarily concerned dispensing battle-field medicine for injured soldiers.[31] Today, humanitarianism impacts norms of state behavior and international institutions. Humanitarian principles arguably introduced the notion of international interventions to end suffering and created initiatives to reevaluate the meaning of state sovereignty.[32] The development of the Responsibility to Protect (R2P), for instance, reflects the expansion of this process, transforming humanitarian concerns from a normative discourse into a foundational element of the international system.[33] In the course of a lifetime, humanitarianism grew from an intersection of religious and cosmopolitan ideals into a gigantic, heterogenous set of agendas, initiatives, and institutions.[34] Nonhumans also played a part in this expansion. However, because of the swiftness of these changes, the organizational scales involved, and the strong emphasis on humanity within humanitarian discourses, nonhuman animals rarely receive scholarly attention or formal analysis. More broadly, nonhuman animals have been omitted from traditional discussions about work, labor, and equity, despite their physical, intellectual, and social capacities play-ing a key part in many allegedly human endeavors. Nonhuman animal labor thus remains a largely unthought ingredient in many discussions of political justice, let alone global political justice.[35] In this regard, several different nonhuman animal species and individual nonhuman animals have made remarkable changes and contributions to humanitarian endeavors. When nonhuman animals are recognized, they frequently become celeb-rity causes or, as the following chapters demonstrate, reinforce strongly paternalistic images of the human–nonhuman animal relationship.[36]

Over the past several decades, the most prominent examples of non-human humanitarians are likely demining dogs. First introduced to the practice of explosive detection in the context of armed conflict, dogs slowly integrated into the land mine and explosive detection units of humanitar-ian organizations. At present, dogs assist as part of land mine clearance operations teams from numerous organizations, including the Marshall

Legacy and the United Nations Mine Action Task Force. They have performed the labor of demining in Cambodia, Columbia, and elsewhere, constituting a highly specialized, globalized nonhuman animal labor force trained to reduce the lethal effects of contemporary warfare.[37] From a few specially trained animals deployed during World War II, hundreds of dogs now work in the employ of humanitarian organizations and receive unprecedented levels of concern, care, and media attention.

Rats are another nonhuman animal with a complex relationship to humanitarianism. Historically, rats have been viewed with disgust and suspicion as the agents of infectious disease and anathema to human well-being. In the past two decades, however, rats have also become contributors to humanitarianism because of their remarkable olfactory sense. Research suggests that rats are more effective at explosive detection in many environments than dogs.[38] Moreover, rats have a notable ability to detect the presence of specific infectious diseases. With these senses cultivated and trained, rats offer new methods of medical diagnoses at a greatly improved speed relative to traditional humanitarian methods of disease management.

Likewise, caprines (goats) and bovines (cows) provide vital services in the creation of sustainable agriculture and play a key role in making milk as a food source and commodity.[39] Nongovernmental organizations like Heifer International and Veterinarians without Borders dedicate a part of their services to the well-being of these nonhumans. Numbering in the tens of thousands, these nonhuman animals have become a bedrock of labor, globally supporting tens of millions of families confronting hunger and poverty. Unlike more highly trained, procured capacities, such as explosive or disease detection, the domestication of these nonhuman animals makes their movement into humanitarianism a new appropriation of their existing relationship with humans. However, even a long-standing practice like dairy production acquires new significance once placed in a humanitarian context.

Many other nonhuman animals participate in humanitarian operations. Cats and dogs may be found in humanitarian compounds along with birds, insects, and reptiles. Nonhumans work as service animals; they perform emotional labor; they are adopted or conscripted into many different living arrangements.[40] Species coming from different places, ecologies, and communities find themselves buried within humanitarian enclaves. Depending on the criteria, anywhere from thousands to millions of nonhuman animals participate in making humanitarian action. These nonhuman animals compose only a small microcosm of the emerging scale of humanitarian logistics because contemporary humanitarian organizations now include dozens of vital resources, hundreds of operations, and thousands of personnel from across the globe; legal entanglements with international institutions, corporate partners, and sponsors; continent-scale relief efforts; and innumerable forms of technical, legal, and professional expertise.

Addressing a small set of agencies, such as nonhuman animals, might seem like a weak foundation for making claims about the politics of humanitarianism, but there are several reasons why exploring the role of nonhuman animals provides unique insights into humanitarianism. First, humanitarianism is not a uniform ideology but a patchwork of developing institutions and practices that appeal to resonant aspirations about the importance of reducing suffering. Frequently, debates about humanitarianism focus solely on questions of the ideal, such as the appropriate circumstances to offer humanitarian aid or the correct conditions to ignore national sovereignty.[41] These discussions have undeniable value, but they often neglect the implicit ideas emerging from the complex patterns of ecological life. Other prominent humanitarian rhetoric sensationalizes, hypes up, or intensifies sympathy in response to human suffering.[42] Yet, none of these discussions elaborates on the division between humans and nonhuman animals when humanitarianism hinges on this distinction for self-definition. Even the earliest appearance of the notion of the "humanitarian" in the context of debates over the status of Christ's divinity or humanity in the late eighteenth century employed foundational distinctions between gods, humans, and other nonhuman animals.[43] The relationship between humans and nonhuman animals is thus a defining element of humanitarian discourse.

Second, humanitarian practices are made up of nonhuman things. Radios, foodstuffs, computer circuits, and caravans are all examples of how nonhuman objects populate humanitarianism. While it is debatable to what degree these objects have the capacity to affect social patterns, nonhuman animals are almost always viewed as possessing some, even minimal, degree of agency and sentience. As this book demonstrates, at first glance, these descriptions may give the impression that only particular, usable nonhuman capacities help promote humanitarian practices. However, exploring these capacities in greater detail reveals that nonhuman animals contribute to humanitarianism in ways that are both complex and singular. Doing so moves the discussion away from a stale, human, all-too-human dialogue about ideal humanitarian virtues and into the messy networks that generate humanitarian practices. Focusing on nonhumans expands existing discussions about intrinsic human empathy or reason that dominate hypothetical explanations of humanitarian politics and, instead, maps the forms of concern, care, and assistance generating their own, largely unrecognized forms of life. Many of the influences of nonhuman animals, as this work documents, have transformative effects on the ecologies and communities they encounter. In this way, the study aligns with a burgeoning literature in international studies on the importance of nonhuman things, ecologies, and interspecies interactions as constitutive of global politics.[44] It also resonates with, while contesting, efforts to rethink political categories surrounding internationalism, cosmopolitanism, and environmentalism on these grounds.[45] Humanitarianism may constitute only a microcosm of international interspecies politics, but it also represents a discourse with a potential to advocate for greater standing for nonhuman entities in international arenas.

Third, exploring nonhuman involvement in humanitarianism provides a tool to reflect on the limits of humanitarian paradigms. Critical analyses of humanitarianism have long demonstrated that it struggles with the category of the inhuman or nonhuman, particularly in the case of aggression and violence. Focusing on flagrantly nonhuman animals that can never make a claim to being human in a traditional sense illustrates the dependence of humanitarian care on underlying conceptions about the human, the animal, and life more generally. By doing so, it offers an opening to consider what about humanitarian practice constitutes an ethical relation to others beyond stipulations regarding the significance of human suffering and well-being, pulling into focus a background in which humans participate in myriad relationships with other nonhuman animals and ecological assemblages. Bringing this background into conversations surrounding humanitarian ethics raises important questions about why humanitarianism attends to human suffering but neglects the suffering and death of nonhuman animals or the destruction of ecologies, especially given the imminence of the sixth mass extinction or the scale and intensity of industrial meat production.

Finally, humanitarian experiments with nonhumans are in many ways far more novel and significant, generating changes at the scale of evolutionary time, than even their most ardent proponents realize. As Marcelo Sánchez-Villagra, director of the Paleontological Institute and Museum at the University of Zurich, comments in reference to rat demining, "this is probably one of the most paradoxical cases of human sophistication—using special skills of other living beings to deal with our own species' proneness to proactive aggression."[46] This comment captures the incredible novelty of nonhuman humanitarians in which other nonhuman animals become a mediating agent that attenuates violence within human communities. Here nonhuman animals are recruited, sometimes using force and coercion, as agents of harm reduction and conflict resolution. This process is paradoxical because it uses multispecies capacities and ecological sensibilities to dampen the intensity of intraspecies conflict. In this sense, nonhuman humanitarianism may be entirely novel as an evolutionary, let alone political, phenomenon. Analyzing the complexities of these changes is thus important not just because of policy, ethical, or even metaphysical questions bound up with humanitarianism but because nonhuman humanitarians constitute a potentially new form of politics offering radical redress to violence through multispecies collaboration.

ON ANTHROPOCENTRIC REASON

Nonhuman labor in humanitarian interventions is already the subject of considerable study.[47] The purpose of these analyses is primarily to discover how best to use the capacities of nonhuman animals for particular ends. These studies pose questions like how to produce the best practices to coax dogs to smell for land mines with efficiency. Here nonhuman animals are framed as a set of tools that perform specific functions and, in many cases, explicitly compared to mechanical or technical devices. This understanding of nonhuman animals is another reason to examine nonhuman humanitarians, because humanitarianism discourses are traditionally anchored in claims about the dignity and intrinsic value of life.[48]

This book uses the term *anthropocentric reason* to describe the mode of thought that characterizes nonhuman animals in most humanitarian operations and epistemologies. Anthropocentric reason understands nonhuman animals as useful or disposable instruments exclusively for human ends. The term shares an affinity with the notion of "instrumental rationality," or a form of reason that reduces a thing to a means for achieving a rationalized end.[49] Anthropocentric reason, by comparison, foregrounds the labor, body, life, and vitality of nonhumans as valuable insofar as they serve discretely human purposes. While it might be sufficient to describe humanitarianism as anthropocentric, humanitarians also develop and deploy a distinctive logic to characterize their relationships to nonhuman animals. There are two elements to this logic: first, to implicitly credit nonhuman animals for specialized forms of labor and, second, to discredit nonhuman animals as fully humanitarian actors. Put differently, anthropocentric reason is worth distinguishing because it rests on a tension between acknowledging the beneficial capacities of nonhuman animals for humanitarian operations and, simultaneously, questioning whether these capacities are humanitarian at all.

Anthropocentric reason enables humanitarian operations to capture or exploit nonhuman animals as a means to an end while publicly condemning the objectification, misuse, and subjugation of human life.[50] However, the integration of nonhuman animals into humanitarian practices according to the logic of anthropocentric reason grounds humanitarianism on an inherently unstable foundation, one that depends on the exclusion of nonhuman animals from political life. As myriad biologists, ecologists, and philosophers demonstrate, the boundaries between microorganism, animal, plant, and human are anything but solid.[51] Human life is unevenly punctuated by organic and inorganic processes and is part of their occurrence.[52] Microbiomes are the stuff of popular science *and* international relations.[53] The ecological crisis has revealed how deeply embedded humans are as part of both planetary-scale ecosystems and niche ecologies.[54] In theoretical terms, the division between human and animal has come under scrutiny for cognitive, biological, psychological, and philosophical reasons.[55] For instance, many nonhuman animals employ complex resonant sign systems, higher cognition, and aesthetic creativity, attributes once assumed to be exclusively human.[56] As Marc Bekoff and Jessica Pierce contend, many "animals seem to have a sense of fairness in that they understand and behave according to implicit rules about who deserves what and when."[57] Nonhuman animals use tools, build territories, anticipate future conflicts and potentialities, augment ecologies, give gifts, and engage in metacommunication in ways that are quite remarkable.[58] Even politics, that domain that Aristotle described as the defining capacity of the human animal, has been thoroughly truncated by nonhuman agents, ranging from the protests of Loukanikos to public debates about Dolly.[59] Nonhuman animals have never been harder to distinguish from human communities, capacities, and lives. Yet, the distinction persists.

Giorgio Agamben offers one explanation of this persistence. According to Agamben, an "anthropological machine" operates within Western politics. This machine perpetually wrestles with the distinction between human and animal because the terms of politics have, since their invention, been caught in an oscillation between these concepts.[60] While Agamben's own work maintains a distinction between human and nonhuman animals, the anthropological machine ceaselessly seeks to articulate this distinction because of an instability or undecidability in its underlying metaphysical terms. As such, over the past two millennia, the machine produces new divisions, new distinctions, between these categories, only to have them collapse into indistinction with one another.[61] For Agamben, this machine functions indefinitely so long as politics remains predicated on an ambivalence about the status of life. Potentially, this ambivalence emerges because of an ongoing process of differentiation or distinction between a thing and itself, which makes any effort finally to establish a secure relation between these terms impossible. Under the sway of this machine, politics becomes an interminable effort to articulate the human–animal distinction to make this separation durable and governable. This process does regularly create seemingly real, resilient, and historically salient distinctions. In some contexts, "the non-man is produced by the humanization of the animal," while in others, it "functions by excluding as not (yet) human an already human being from itself, that is, by animalizing the human, by isolating the nonhuman within the human."[62] In both cases, the anthropological machine seeks to determine the distinction between human and animal and extract from this distinction a foundation for politics. Although Agamben's account of nonhuman animals has notable problems, the notion of the anthropological machine explains why the relations and thresholds between humans and nonhuman animals are so crucial to the articulation of contemporary political experience.[63]

At this historical moment, as scientific discovery, psychological insight, and philosophical thought question the salience of human–nonhuman distinctions, humanitarianism has both integrated nonhuman animals into its practices and rearticulated the division between human and animal. In effect, humanitarianism both "humanizes animals," by making nonhuman animals into humanitarian actors, and "animalizes the human," by capturing human suffering as what Agamben terms "bare life." In this way, humanitarianism has become a new site for the anthropocentric machine to temporarily reestablish the caesura between human and nonhuman. Within humanitarian discourses, ethics, care, compassion, and reason operate as a set of concepts, shifting semiotics, frequently displacing one another, that craft actionable but impermanent, durable but delicate, distinctions between humans and nonhumans. Each example of humanitarian involvement with nonhumans enacts this in slightly different ways, and so actual distinctions between humans and animals, human and nonhuman humanitarians, are never articulated but always in a process of articulation. In this sense, while previous regimes of anthropocentrism discredited nonhuman capacity for ethical action, humanitarianism offers a particular blend of enthusiastic affect, compassionate concern, and faith in the comprehensiveness of human faculties to offer a more elaborate, heartfelt, and sanguine version of the difference between humans and nonhumans. Humanitarian organizations also sometimes celebrate and elevate nonhuman animals as capable of humanitarian virtue while simultaneously presenting forms of human life that are endangered and exposed as reduced to a condition akin to base animality. Ultimately, animality is the form of otherness that humanitarianism seeks to save humans from, and so humanitarian discourses are bound to return to reinforce this distinction. In doing so, humanitarianism perpetuates the anthropological machine, articulating but also disarticulating human and animal. With each distinction, a novel formation of violence and exclusion (but also the possibility of their contestation) emerges.

Two aspects of humanitarianism are particularly important with respect to this point. First, humanitarianism shifts the underlying terms of emphasis so that an ever-changing combination of reason, empathy, and outward expression is necessary to achieve humanitarian virtue. This conceptual mobility is why humanitarian advocates constantly debate the "origins" or "bases" of humanitarian action only to arrive at arbitrary or contingent, if politically potent, formulations. Second, the orientation of humanitarianism, toward the improvement of human life, establishes a loop where simply possessing the capacity to act in a caring or compassionate manner is not sufficient to be humanitarian. Rather, these capacities must also be inclined or directed toward the improvement of human life a priori. This structure creates a bind for nonhuman animals. On one level, nonhuman animals must prove that they are capacious in humanitarian arenas, but on another level, they must prove that these capacities emerged to improve human life.[64] By explicitly defining compassion and care as distinctively human features, but implicitly treating these capacities as meaningful only when oriented toward the improvement of human welfare, it becomes impossible, no matter how many capacities, virtues, or generosities nonhuman animals display, for a nonhuman animal to become a humanitarian. If a nonhuman animal shows the proper capacities, then it fails to show the appropriate orientation. If a nonhuman animal displays the proper orientation, then it fails to have this orientation built a priori into its emotive capacities for care. However, even if a nonhuman demonstrates both propensities by possessing the right faculties and the proper emotional orientation, because it has been bred and trained for this purpose, this demonstration problematizes the boundaries of the human so that the very act of proving that an animal can be, in a sense, fully humanitarian undermines the cogency, and therefore the protection afforded by, the status of being human or humanlike. In this move, humanitarianism establishes the human as a vanishing point that perpetually discredits the capacities of nonhuman animals or, alternately, dissipates once nonhuman animals appeal to humanity as a meaningful category of political protection.[65]

Following this model, humanitarianism elevates the importance of specific forms of human suffering as an object of modern power and acquires a new importance in contemporary politics as a means of resuscitating the anthropological machine. Indeed, the present is replete with challenges to humanitarianism, including the resurgence of xenophobic nationalism and fascism, the growing importance of economic imperatives, and the multiple ways power politics, racism, and the colonial past truncate humanitarian objectives. Humanitarianism persists partly because the model of suffering, identity, and ethics it supplies does not jeopardize all these power structures as much as it calls for a normative shift to credit and care for humans because of their inborn propensity for virtue. In doing so, humanitarianism not only becomes an artifact of global power but enables an implicit negotiation over the terms of the human. Anthropocentric reason works as a counterpart to this process. It explains why nonhuman animals are always lacking some capacity, some characteristic, that bars them from being defined as humanitarian agents but nonetheless allows humanitarian organizations to capture their capacities to strengthen humanitarian practices. By discursively isolating humans as the affective and ethical agents of humanitarian practices, it becomes more difficult to envision nonhuman animals as capable of forming reciprocal, caring, political relations with others independent of human governance or supervision. When nonhumans do forge caring relations, this response is interpreted as base instinct rather than a reciprocal, relational, or dynamic form of engagement involving risk and uncertainty, let alone a politics.[66] Put differently, anthropocentric reason is how humanitarianism domesticates the potential of nonhumans, defangs this potential in the form of a cute appreciation for animal joviality, and milks it for the health of specific human lives. However, the notion of anthropocentric power is also mutually constituted with the destruction and genocide of social worlds within more traditionally human politics. Here anthropocentric reason helps illustrate the specific apprehension of nonhuman animals within humanitarianism even as anthropocentrism and the Anthropocene are historically bound up with various forms of discriminatory politics.[67] While not disconnected from these larger questions of human equity, here the point is to highlight the logic that informs the movement of nonhuman animals into the service of humanitarian institutions as well as the way humanitarian organizations implicitly disempower and animalize the victims they seek to serve.

Stepping back from this more theoretical point, anthropocentric categories also generate inequities for the nominally human subjects of humanitarian discourse. Scholars have noted that humanitarianism often tacitly relies on the construction of inhuman others to structure its ethical comparisons.[68] Representations of savages, victims, and saviors abound within humanitarian spheres. Critical theorists, scholars of race, gender, and other axes of domination, have demonstrated that the human is a category created for political purposes rather than a natural phenomenon.[69] Postcolonial writers have exposed the limits of humanitarianism with respect to the colonial condition and explored the non-Western origins of many humanitarian ideals.[70] Many scholars have documented the many compromises, contradictions, and constraints of humanitarian operations as well as their capacity to expand rather than curb violence.[71] Analysts of race and anti-Blackness have brilliantly illustrated how the gratuitous violence of slavery and its afterlife was a prerequisite to the invention of humanism and the development of humanitarianism.[72]

These insights demonstrate that ascribing humanity is not a benign observation. Rather, the concept of the human establishes a hierarchy between different forms of life. It facilitates acts of ontological policing over the boundaries of humanity, determining what identities, expressions, and bodies should exist and for what purposes.[73] In doing so, humanitarianism also produces its others, those forms of life that are either insufficiently human or antithetical to the human condition. Race, gender, sexuality, intellect, disability, and animality all historically generate hierarchies within contemporary human communities and, by extension, exclude forms of life that can never make a claim to humanity in the first place. In a sense, the human operates as a regulatory zone that is at times violently defended and at times lazily adjudicated using both epistemological and material practices. This zone actively filters, suppresses, subjugates, and eliminates different modes of life. It is impossible to offer a postracial, ungendered, unsexed, acultural, or apolitical account of the human. The concept operates as a means of both establishing racist hierarchies and securing the terrain for their negotiation.[74] It is a term that naturalizes specific models of gender and sexual difference.[75] It is an idiom that implies that a properly human body should possess specific cognitive and physical features.[76] Ironically, nonhuman animals, who have always struggled to enter into negotiations over the category of the human because of cognitive, sensory, and morphological differences, are often employed to strengthen the salience of these comparisons. Consider, for instance, the colonial juxtaposition of Black bodies with chimpanzees and apes as a method of displaying the continuities between humans and animals.[77] These displays, equal part performances of colonial spectacle, white supremacist entertainment, and mechanisms of knowledge production, produced the meaning of human, animal, and quasi-human through these rituals, or consider the way that dogs both produced and secured the ontological boundaries of the plantation by tracking and hunting runaway slaves.[78] Alternately, consider the use of dog breeding as a marker of gendered, raced, and classed privilege or the not infrequent inequitable comparisons made between the cognitive faculties of nonhuman animals and neurodiverse humans.[79] Species difference, as Donna Haraway says, stinks of sex, race, and ableism.[80] Nonhuman animals, as creatures that struggle to make any claim on the human, historically function as a perverse complement to the category of the human as laborer, as regulator, and as a touchstone securing the epistemological foundations of the distinctions. As Bénédicte Boisseron points out, violence against nonhumans is connected or assembled with these other atrocities, resonating, sharing some features, while also constituting distinct modes of violation.

The exclusionary potential of humanitarianism, its participation in structures of racial, gendered, ableist, and colonial power, despite its universalist aspirations, is a well-documented, if contested, point. Yet, this exclusionary potential has rarely been thought in relation to nonhuman animals. This oversight is strange precisely because the human and its others are directly at stake in the self-definition of humanitarianism.

Focusing on the role of anthropocentric reason and charting the place of nonhuman animals in humanitarian politics adds four insights to the critical dialogue surrounding humanitarianism. First, it foregrounds the way that humanitarian operations are always already more than human endeavors. Debates about the ethical merits of humanitarianism typically advance claims based on the universality of human value and the possibility of human reason and empathy, only to encounter normative crises or inconsistencies that undermine these ethical visions.[81] Didier Fassin, for example, shows that a permutation of empathy and rationality is constitutive of humanitarian governance.[82] From the vantage point of nonhuman actors, it is the distinctive emphasis on *humanity* as political subject and *human* faculties as a political instrument that facilitates responses to violence and deprivation. This raises the question of why reason and empathy are understood as specifically human traits. What is at stake in assigning the status of the human to these features within the field of humanitarianism? The response many humanitarians implicitly offer to this question highlights how human faculties, human capacities for connection, are the reasons humans value one another. Humanitarians claim to celebrate not just the mere fact of being human but the distinguishing features of humans as the site of ethical responsibility.[83] In short, it is what humans do, how they reason and feel, rather than simply being human, that makes humanitarianism what it is.

However, this response only deepens the problem in relation to nonhuman animals. When compared in terms of their capacities, biologists, geneticists, and neuroscientists have documented few, if any, distinctive traits of humans except for sweating and throwing.[84] Neither of these features strengthens the claim that humanitarian sentiments uniquely define humanity. Rather, they help to reveal the role of fantasy in structuring humanitarianism. This fantasy endows the thing called human, a supposedly stable referent, with a special character, which widens the gap between humans and nonhumans to produce and sustain hierarchies among those who can make a claim to the human. By adopting this language, humanitarianism tacitly participates in a dialogue on nonhumanity with a twisted history that both generated the horizon of democratic freedom and constituted the basis for slavery, war, and genocide. Invoking the value of humanity does not ensure that one's cause enhances freedom, alleviates suffering, or improves human welfare. Despite aspirations to the contrary, claiming the figure of the human *guarantees* virtually no meaningful outcomes in contemporary politics. Atrocities, violence, structural deprivation, are not simply the product of a failure to recognize the humanity of others but outcomes that take the concept of the human as their condition of possibility. However, as Peter Redfield notes, "the field of humanitarian concern is clearly focused on a fluid and expansive conception of vital need, spread beyond the citizen to the figure of the human."[85] Nonhuman animal labor in humanitarian contexts raises the questions of whether this field of concern can be extended to include other species or forms of life and what consequences would result from such an extension.

Second, in contrast to humanitarian ideals, which center directly on human dignity and suffering, actual existing humanitarian practices are composed of myriad nonhumans. Although this work focuses primarily on the role of nonhuman animals that contribute to humanitarian practices, humanitarian operations involve all manner of nonhuman things, including tents, fiber optics, computers, medical kits, telephones, foodstuffs, automobiles, compounds, instructional manuals, stone, GPS, donations, and advertisements.[86] Though these things are also produced by social discourses, any viable humanitarian operation would be a nonstarter without their material capacities. Nonhuman animals play increasingly complex roles in humanitarian operations. Most of these functions are either invisible, a part of the background of humanitarian activity, or legible solely through anthropocentric reason as nonhuman animals become specialized tools, their capacities bred, trained, and selected to further a specific objective defined by humanitarians. Paradoxically, the humanitarian emphasis on universality, on the implicit value of human, exposes nonhuman lives to death. It is a form of universalism predicated on the exception.[87] Foregrounding nonhuman humanitarians not only highlights the anthropocentric violence that accompanies humanitarianism but demonstrates how profoundly nonhuman humanitarians have reworked, expanded, and transformed humanitarianism. This shift enables a redefinition of humanitarianism away from generalizing tropes about the human and toward entangled practices of care that work both within and beyond human, all-too-human imaginaries.[88] The point is not just to offer a polemical critique of the limits of humanitarianism but to seek within humanitarianism, despite the problematic way it articulates nonhumanity, the occurrence and potential for new means of contesting violence and the possibility of pluralist, generous forms of political engagement that span communal and species difference. Indeed, this study takes to heart that humanitarianism is an effort that should be greeted with what William Connolly calls "presumptive generosity" and "agonistic respect."[89] These dispositions involve showing the contestability of the underlying faith humanitarian organizations maintain in humanitarian principles and the human–animal distinction while also remaining open to unforeseen encounters and possibilities this work creates. Humanitarian organizations are striving to produce a form of flourishing, and although frequently problematic, these efforts should also be credited with generating new forms of life and possibility. This disposition also requires understanding humanitarian practices and actors not as discrete, dignified, and autonomous subjects but as plural, deeply relational, and more than human.

Third, exploring anthropocentric reason also deepens existing critiques of humanitarianism by highlighting what is arguably their most fundamental exclusion: forms of life that could never make a claim to the category of the human in the first place. Dogs, camels, goats, and rats may all constitute companion or commensal species and have, to varying degrees, lengthy coevolutionary histories with humans.[90] They do not, however, remotely resemble the forms of life envisioned by the category of humanity. Consequently, a politics derived from an emphasis on human suffering will by default ignore the concerns and considerations of these species. As Anna Tsing argues, the all-too-human generates economic and ecological frameworks that invite ruination for nonhuman landscapes and all too easily naturalize the exploitation of nonhuman others.[91] Although the majority of this study is concerned with the implications of this exploitation in the context of humanitarianism, in the background lurk broader questions about the value of focusing on the question of human life and well-being in an age characterized by unprecedented mass extinction, forms of consumption that require the habitual reproduction and destruction of multiple species, and anthropogenic climate change. In terms of ethics, this point raises questions about the value of what humanitarianism foregrounds when the background is arguably defined by gratuitous destruction (both deliberate and unintentional) of countless nonhuman life-forms. It also raises important questions about how nonhuman animals reinforce, resist, or disengage with the multiple anthropocentric forces they encounter in humanitarian contexts.[92]

Fourth, by providing a new lens into the role of care in the politics of humanitarianism, nonhuman humanitarians open insights into the ways that humanitarian operations benefit vulnerable human and nonhuman communities. As Brian Massumi argues, becoming animal involves dynamic forms of play and creativity.[93] In attending only to the instrumental capacities of nonhumans, humanitarianism ignores the significance of nonhuman care in both theoretical and practical ways. At a theoretical level, nonhuman competencies reveal a need to reassess whether it is the human as agent of care or care as an emergent, relational capacity of humans and nonhumans that defines the future of humanitarianism. At the level of practices, many humanitarian interventions that incorporate and rely on nonhuman animals often involve unforeseen benefits to the recipients of humanitarian services, including affective labor, such as soothing, comforting, and forging bonds, that would be impossible if strictly defined by human face-to-face encounters. As such, revealing a more complex human and nonhuman humanitarianism is a means of revisiting what makes humanitarianism worthwhile as a form of politics.

Oddly, humanitarianism, despite foregrounding the human, has rarely been evaluated in relation to anthropocentrism. Even critical accounts of humanitarianism often challenge the reliance on an implicit concept of the human only ultimately to seek to expand the scope of the term by advocating for a broader recognition of more experiences, more models of human life. This is a valuable gesture, but it leaves the anthropocentrism of humanitarianism intact. Highlighting the role of humanitarianism thus offers a mechanism of deepening and extending previous critiques by showing how thoroughly debates surrounding humanitarianism remain bound to a strong notion of a human-centered universe.[94] Calling out this function is part of what is at stake in thinking about humanitarianism in relation to anthropocentric reason. In contrast, at multiple points, this text discusses features or capacities of nonhuman animals considered valuable to humanitarian interventions. It makes these points partly to trace the systems of thought that incorporated nonhuman animals into humanitarian projects but also to comment on the way that different nonhuman animals (or other forms of life) participate in an ecological context in which they coexist with humans. This argumentative strategy engages in anthropomorphism because it requires describing different models for how nonhumans might engage their reality largely through secondary effects. Unlike anthropocentrism, which makes implicit normative commitments about the value of human perspectives and life, anthropomorphism results from the way humans access, engage, or perceive reality. There is no "outside" to these filters because a thing is constituted by its mode of sensing its reality.[95] The inevitability of anthropomorphism means any analysis should be cautious as it makes speculative claims about the makeup of nonhuman worlds, because these can easily support reductive, anthropocentric fantasies. Throughout the text, I attempt to mark this move into a speculative discussion of the incommensurate encounters between humans and nonhumans or nonhumans and other nonhumans. By doing so, I hope to consider the ethical possibilities of these encounters while avoiding a genre that turns nonhuman animals into a pantomime of human experience.

ANTHROPOCENTRIC FEELING, HUMANITARIAN SENTIMENTS, AND AFFECTIVE LURES

Even with these qualifications, there are important dangers associated with exploring the possibility that nonhumans participate in a humanitarian politics of care. First, doing so risks positioning generosity as a base animal feature and excluding animals that fail to reflect this norm. Second, studies of nonhuman animals often focus on nonhumans that most closely resemble or best approximate highly normative, humanized forms of sociality: fluffy, vertebrate, engaged mammals with an evolved, conditioned, or domesticated aptitude for the forms of coexistence most admired by humans—in other words, cute, tame, polite nonhumans that display features like neoteny rather than the ugly, hazardous, or wild.[96] In this way, cuddly, docile, and comforting critters become objects of paternalistic affection.[97] As Donna Haraway argues, "resistance to human exceptionalism *requires* resistance to the humanization of our partners."[98] Moreover, producing nonhumans as objects of affection arguably operates as a complement to practices of slaughter by differentiating cute animal from consumed meat.[99] Sociability thus needs to be treated as a political artifact as opposed to an inherently productive tendency toward bonding or mutuality. Finally, appreciating nonhuman care chances a kind of anthropocentrism by concentrating solely on the attributes of nonhumans most favorable to humans and sneaking in another version of anthropocentrism. These concerns are especially pertinent in this context because humanitarian discourses have demonstrated a consistent ability to deflect, marginalize, or suborn criticisms through an appeal to feelings, emotions, and sentiment.[100]

In their overview of the historical development of humanitarian institutions, Michael Barnett and Thomas Weiss confess that "it is difficult to avoid writing a highly sympathetic, nearly sycophantic account of humanitarianism or interpreting its evolution and expansion as a sign of moral progress."[101] Many state officials, professional humanitarians, and public intellectuals appear to agree and openly embrace humanitarian virtues.[102] What is more surprising is the number of critical intellectuals who make the case for humanitarianism despite their awareness of its more problematic dynamics. Lynn Hunt, for instance, forcefully argues that human rights and humanitarianism emerged from epistolary novels and the exposure to an imagined world of the other. Cultural forces thus account for the rise of humanitarianism despite the latter's ambivalence about suffering in cases ranging from gender-based discrimination to chattel slavery.[103] Kathryn Sikkink's reading of the "justice cascade" similarly outlines a progressive account of the rise of justice-oriented and humanitarian international institutions.[104] While Sikkink famously described the many compromises, stagnations, and complexities of global advocacy networks, the ideals associated with humanitarian vision appear to override these intricacies.[105] Many activists and scholars from marginalized communities also embrace, if situationally, humanitarian outreach as a kind of "politics from below" that may improve contemporary social conditions.[106] Even philosophers who situate themselves far afield from humanitarian traditions, such as Jean-Luc Nancy, occasionally find themselves swept up in the momentum of humanitarian declarations.[107]

The resonance of humanitarian sentiments with scholars and communities that identify multiple dangers in humanitarian practice indicates the presence of an ideology, one that adheres to Slavoj Žižek's formula "I know very well, but . . ." In this case, the logic goes something like this: "humanitarianism may be colonial, hypocritical, inconsistent, reflect power politics, white supremacy, opportunism, or Eurocentrism, but I still believe in its possibility." David Kennedy's *The Dark Side of Virtue* offers a potential example. Kennedy's work chronicles the many faults of humanitarianism, primarily the way it has become a mode of governance that embraces problematic, violent practices, only ultimately to articulate a "hope [that the] humanitarian impulse can be made real in the world."[108] Here Kennedy holds out on the possibility that activists and policy makers can renew humanitarianism with new principles and emphases and, because of this possibility, ends up embracing humanitarianism despite his other objections.[109] This aspiration, the hope that humanitarianism could be otherwise, that it could more meaningfully help suffering others, is a key part of its ideological makeup. In the context of this study, it is important to acknowledge this tendency to resist an analytical temptation to interpret nonhuman humanitarianism as indicative of a new, better version of humanitarian politics rather than a deeply contested phenomenon involving both power and ruses, creativity and exploitation, harm and solidarity. Indeed, the prevalence of scholars articulating the virtues of humanitarianism raises the question of how humanitarianism became a dominant ideological prism for understanding problems like mass violence and suffering. What about humanitarianism makes its rhetoric so successful as a lure that statespersons, youthful activists, and critical intellectuals resonate with its aspirations despite their various reservations? How does humanitarianism remain influential for so many communities despite their misgivings about its virtues and effects? The answers to these questions might further clarify what roles nonhuman animals play in the production, promotion, subversion, and transformation of humanitarian politics.

Several scholars have provided potential responses to the question of which social forces led to the rise of humanitarianism. Samuel Moyn brilliantly illustrates that human rights and humanitarianism constitute a "last utopia" or set of minimal imagined political freedoms that began to coalesce in the 1970s. During this period, the decline of liberal-democratic capitalist ideals, the dissolution of self-determination movements, and the disintegration of grand-scale communism enabled human rights and humanitarianism to emerge as a shared idiom for conceptualizing freedom absent a broader ability to change the world.[110] Put differently, human rights replace a more expansive, if faltering, conception of international justice. In doing so, they create the minimal promise of meaning within politics. Humanitarianism helps fulfill this promise because it maintains the possibility of a politics that fights against the worst. Jessica Whyte, in contrast, argues that humanitarianism and human rights were seized and transformed by neoliberal economic forces seeking to develop a model of rights that could accommodate a market-driven society. Conceptions of human justice adopted a form that resonated with a specific model of capitalist economic administration.[111] By way of comparison, Costas Douzinas contends that humanitarianism operates according to the logic of the Lacanian symbolic, a stand-in for a lack in the big Other that constantly bombards us with instructions to act on behalf of "the human" (no matter how vacuous or how vulgar that human might seem to be). For Douzinas, the complement of this injunction to consider the well-being of others is a narcissistic enjoyment and superficial distance that such concern often enables. He alleges that it is this very distance that makes humanitarianism an ideal doctrine for contemporary imperialism.[112] Stephen Hopgood views humanitarianism as predicated on a unipolar moment of Anglo-cum-American power. In this account, humanitarianism depends on a system of international governance that no longer truly functions as it did in previous historical moments because it relies on these highly limited sets of political conditions.[113] Modern attachments to humanitarianism reflect a dying effort to reproduce an unfortunately outmoded set of international regimes. Sharon Sliwinski approaches the problem by tracing the aesthetics of humanitarianism to descriptions of the 1755 Lisbon earthquake, which popularized images of extreme human precarity.[114] In her account, the aesthetic practices that birthed humanitarian sentiments arose from the dissatisfaction with explanations derived from theological sources. Unlike narratives of divine disaster, humanitarian stories are always stories about our supposedly shared condition. These reflections on the plight of other humans paradoxically become reflections on our own plights, thereby generating a loop in which concern for the other is always already also concern about ourselves. Humanitarianism works because it is a story about others that is also always already about ourselves.

These responses offer a sample of accounts that explain the genesis of humanitarianism in relation to multiple historical factors, narrative procedures, and social structures. They show how humanitarianism resonates with different creeds, traditions, and beliefs surrounding care and ethics, images of intense violence, and the possibility of meaningful connection to others in an age characterized by logistically designed disconnection, global mediation through digital space, counterrevolutionary forces, and growing planetary risks.[115] They situate humanitarianism as a more minimalist response to alternative models of utopian political action and challenges to global inequity. Put simply, they document how many different anchors support the growth and persistence of humanitarianism. What these arguments also demonstrate is that humanitarianism is defined not solely by one set of principles but by a series of institutions, practices, expressions of interest, advertisements, and pedagogies. Given this multiplicity of different sources, it might be useful to read humanitarianism as what Gilles Deleuze and Félix Guattari call an "apparatus of capture."[116] An apparatus of capture is an assemblage that seizes on an existing set of heterogenous social and ecological differences, redirecting flows, movements, traditions, and practices, and recodes them for new purposes.[117] The apparatus is an abstract phenomenon. It exerts pressure on ideas, discoveries, and ways of life, causing them to resonate with one another and governing them according to a new set of imperatives. An apparatus encourages specific tendencies by crafting axioms, sets of articulated and unarticulated rules, that ensure isomorphoric relations across the different areas governed by the apparatus.

The various dimensions of humanitarianism described by Moyn, Whyte, Douzinas, Hopgood, Sliwinski, and others are examples of axioms that hold humanitarianism together. They offer different points of origin where humanitarianism resonates because of the way it intersects and changes understandings of politics, freedom, aesthetics, justice, religion, and morality. However, understood as an apparatus of capture, humanitarianism also exceeds any single point of origin and becomes capable of quasi-autonomous development. Indeed, Deleuze and Guattari argue that an apparatus of capture primarily influences tendencies in already existing social formations; it operates across contexts without being immediately apparent in any of them.[118] Over time, the apparatus becomes self-forming because it reconstitutes, and in doing so, captures, a set of flows, capacities, or identities. Humanitarianism closely follows this model. It employs a variety of existing practices within medicine, logistics, engineering, aesthetics, and so on, converting them to new ends, while also resonating with existing doctrines of both secular and religious faith about the importance of caring for other humans. Most important for the purposes of this work, interpreting humanitarianism as an apparatus helps to explain its encounters with nonhuman difference across a variety of contexts. In each, humanitarianism both materially captures the capacities of nonhuman animals and the relations they form to human communities, other life-forms, and diverse ecologies and repurposes these in alignment with the tendencies for humanitarian projects.[119] In this sense, humanitarian relations with nonhuman animals bear remarkable similarity to other techniques of governance. Humanitarianism generally displays a propensity to treat all disasters facing humans as isomorphic in character and as integral to the humanitarian project. Hence, as humanitarianism expands to include not just the care of wounded soldiers on the battlefield but migration, famine, drought, poverty, war, terrorism, and a host of other events, it recodes these events into an isomorphic set of crises that can be redressed through similar processes of aid, assistance, intervention, and emergency management. In doing so, humanitarianism draws on a wide variety of tools, each of which becomes an object of knowledge and practice. Nonhuman animal capacities constitute another dimension of this process as they address preexisting human suffering while bolstering a discourse on the innocence of humanitarian labors. Indeed, as Deleuze and Guattari point out, an apparatus of capture works by making "the mutilation, and even death, come first" as the object to which it responds.[120] Humanitarianism operates not only to alleviate human harm but *by seeking to govern how the alleviation of harm takes place.* As an apparatus, humanitarianism employs nonhuman labor alongside a wide range of international institutions, media campaigns, religious outreach groups, everyday political vocabularies, aesthetics of pain and suffering, and various technologies to create a potent paradigm of contemporary global governance.

An apparatus of capture is neither inherently good nor inherently bad but changes the tendencies of a given process. Engaging the different mechanisms and axioms of humanitarianism is a key part of unpacking its ideological force because there is no single core or foundational source of humanitarianism but rather a cluster of overlapping, dynamic reinforcing beliefs and practices. This creates a problem: humanitarianism cannot be changed solely through recourse to criticism because it is capable of evolving, abandoning axioms and points of emphasis, in the process of reinventing itself. Overtly cynical critiques that claim to dismantle the power structures surrounding humanitarianism typically only criticize specific axioms or pieces of humanitarianism while leaving others intact. Consequently, humanitarianism mutates in response to a critique, abandoning specific axioms, but without ever changing the underlying dynamics of power.[121]

Emotion or affect constitutes a particularly important and underanalyzed component of ideology.[122] Affective lures are part of what produces humanitarianism as an apparatus of capture. For example, consider Barnett and Weiss's description of their lingering sympathies for humanitarian causes quoted earlier, despite their reservations about humanitarian politics. By producing affective resonances with humanitarianism, normalizing these as features of contemporary subjectivity, humanitarianism becomes resilient. A mode of critique that highlights theoretical or intellectual tensions within humanitarianism while overlooking these sentimental connections misses how affect renews investments in humanitarian promise. As humanitarianism subsequently reforms, abandoning certain axioms and adopting others, it strengthens this investment because it demonstrates humanitarianism's capacity to change as proof of its promise. Many of these tendencies are evident in the underlying premises about humanitarian action. Humanitarianism thrives from the strange assertion of contraries that help to articulate the importance of feelings for its institutional coherence. For instance, humanitarians express a belief that an individual's budding concern for the other's well-being can serve as the basis for eliminating planetary-scale injustices; or that empathy and imagery, pulsed though the channels of a global communications network, will remedy the catastrophic consequences of historically specific mass atrocities; or that performances of respect for the other constitute, no matter how removed, a mode of substantive activism. The imagined connection with others and the political status afforded to individual sentiment are examples of axioms that reinforce humanitarianism by normalizing a set of emotional reactions as both fundamental to one's status as a human and vital to disinterring structures of political power. It is a compelling process because it transforms internal sentiments into a statement of a subject's moral character, a means of feeling a sense of possessing political agency, and a mode of social justice. Put differently, humanitarianism makes affective reactions central to its enactment. These reactions help to explain the persistent attachment to humanitarianism despite its various issues.

The affective aspects of humanitarianism require special attention in this case because they also resonate with what is called *anthropocentric feeling,* or a set of sympathetic affiliations and emotional similarities presumed to exist between humans and nonhuman animals. Anthropocentric feeling might be read as anthropocentric reason's complement. Where anthropocentric reason emphasizes the utility of nonhuman animal capacities, anthropocentric feeling takes pleasure in nonhuman emotion and affect that mimic human aesthetic expectations. For instance, human perceptions of nonhuman animals are often filtered through emotional genres that emphasize the suffering, innocence, and vulnerability of nonhuman animals to demonstrate their kinship, affiliation, and linkages with humans and to advocate for their protection and welfare.[123] This affective genre is common in humanitarian materials, and it often forms a synthetic part of the otherwise instrumental emphasis of anthropocentric reason. Moreover, anthropocentric feeling and humanitarian sentiment have resonant political grammars, narrative structures, and emotional genres. The suffering but potentially happy animal easily trades place with the innocent child.[124] Any exploration of the role of nonhuman animals in humanitarian projects needs to be doubly cautious because the affective lures of humanitarianism support reductive versions of anthropocentric feeling, and vice versa. In addition, both models of sentiment are concerning because they also participate in a fantasy that emotional reaction is the key to cutting through power, violence, bureaucracy, and social distance. The lure of humanitarianism (and its ability to capture) relies partly on this fantasy and the virtues it promises.[125] That this promise is frequently cliché, coded, raced, sexed, and overtly melodramatic does not matter because it hints at a possibility of intimacy to the other's vulnerability that drives the entire process, a possibility for intimacy that also enchants human relationships with nonhuman animals.[126] These fantasies make it difficult, if not impossible, to contend with other permutations of political possibility, other forms of care, other ways of attending to the traumatic residues of violence.

As mentioned, the concern here is that celebrating caring relations between humans and nonhumans simply merges anthropocentric feeling with humanitarian sympathy. In anthropocentric feeling, nonhuman animals are understood through emotional genres that emphasize their cuteness, near-human sentiments, or evolved capacity for affinity. Even highly theoretical accounts of nonhuman animals, such as Levinas's reading of Bobby, indulge in this emotional genre, appreciating the affective engagement of nonhumans and viewing them with sympathy, while simultaneously delegitimizing nonhuman animals as deficient with respect to other capacities, such as moral reasoning.[127] Causes centered on animal welfare, animal rights, and animal well-being often appeal to similar frameworks oriented around the illegitimacy of pain and suffering and the innocence of nonhuman animals, concerns deeply resonant with humanitarianism, as writers dating back to Jeremy Bentham demonstrate.[128] Ultimately, the politics of humanitarian sentimentality and anthropocentric feeling participate in what Lauren Berlant calls "cruel optimism." Cruel optimism emerges whenever a situation, event, or condition promises a different future while undermining the capacity to flourish. With cruel optimism, "desire is memorable only when it reaches toward something to which it can attach itself; and the scene of this aspiration must be in relation to the repetition of another scene."[129] In this context, this aspiration defines the promise of humanitarianism and justifies toil in the hope that practices based on empathy, reason, and humanism will reduce human suffering. In the context of anthropocentric feeling, cruel optimism subsists in the expectation of a certain kind of fulfillment that develops from proximity to nonhuman animals, which can address gaps or fits in emotional or political well-being. These situations involve a structure of desire that *wishes* that specific feelings *could* relieve conditions of privation and violence. This wish implies "that structures and institutions of power can always be overcome by personal feelings, personal choices," but, in the end, these attachments persist in undermining flourishing.[130] Consider, once again, Barnett and Weiss's argument that humanitarianism is a type of power relation, but one that *feels important* even to its critics because of its sympathetic agenda and ideals. These sentiments reflect the way that "fantasy donates a sense of affective coherence to what is incoherent and contradictory"[131]—in the context of this study, the way that anthropocentric feeling donates a sense of emotional coherence to humanitarian work with nonhuman animals while simultaneously capturing, exploiting, and, in some cases, killing these very nonhumans in the pursuit of humanitarian possibility.

Put differently, "everyone knows" that humanitarianism does not always produce immediate relief for those facing the greatest plights, that it can be politicized, truncated by bureaucracy, and subject to power, an outgrowth of the colonial dynamics, or an opportunity for corporate tyrannies. Moreover, "everyone knows" that our feelings, no matter how altruistic, do not unilaterally dismantle power or translate into an unmediated access to others. Nonetheless, we have the sense that humanitarian aspirations might be productive, valuable, and even inviolable at their core. Because of this affective structure, simply documenting the power structures at work in humanitarianism cannot, by itself, produce change. In part, the very ambiguities highlighted by these critiques only strengthen the affective lure of humanitarianism because they create the impression that, underneath or alongside these existing practices, there exists the possibility of a better, unmediated remedy for helping the suffering of the other. These fantasies sometimes support terrifying declarations about the ethical character of humanitarian violence. Calculations of this type make apolitical declarations about the necessity of accepting the lesser evil and reveal the underside and violence of humanitarianism fantasies.[132] Simply adding anthropocentrism to the list of hypothetical critiques of humanitarianism is insufficient as a strategy for engaging the problematic aspects of humanitarian politics because the emotional genres of anthropocentric feeling are primed and ready to rebuild humanitarianism on new axioms.

However, there is also no obvious way to avoid this problem. Simply wishing away sentiments is impossible if they are constitutive of institutional relationships and subjectivity. Throughout this analysis, the goal of exploring nonhuman humanitarians is neither to recuperate humanitarian possibility nor to reject it outright. This book does not argue that the cuteness, docility, and companionable character of specific nonhuman animals offer a truer, better, more meaningful, or more beautiful version of humanitarianism. Neither human nor nonhuman animals are, under it all, the "real" humanitarians. Rather, the argument unearths the awkward, entangled, provisional forms humans and nonhumans encounter, their disparate, ecological entanglements, that constitute the conditions of possibility for the emergence of specific forms of generosity, care, and politics ongoing in these humanitarian labors. In doing so, it seeks, not to dismiss the emotive aspects of encounters with humans or nonhuman animals, because both matter, but to demonstrate that these do not necessarily need to be understood in either humanitarian or anthropocentric genres, nor does it articulate an expectation that the existence of a small-scale engagement with a politics brewing at the edges of human and nonhuman worlds fundamentally changes any global political arrangement. The point is more modest.

Each chapter explores the underlying ecological capacities that enable nonhumans to create relations of care with one another, humans, and broader milieus. Each chapter traces the institutionalization of these relations within humanitarianism and illustrates how fantasies of sentiment and compassion normalize these connections and expose certain forms of nonhuman life to death and violence. At the same time, the chapters seek to denaturalize these types of care, to display the weirdness and divergence found in human–nonhuman coexistence, and to demonstrate that anthropocentric reason and anthropocentric feeling are reductive genres for interpreting nonhuman humanitarians. In effect, the book argues that, within humanitarian institutions and interventions, a multiplicity of experiments in multispecies justice including different subjects, models of ethics, modes of communication, and forms of solidarity are breaking with the normative expectations of humanitarianism.[133] Put differently, the book subtracts anthropocentric value, sentiment, and emphasis to reveal plural forms of human–nonhuman collaboration that tackle different problems of political justice and the ethical challenges of coexistence in global politics. Although it is true that some of these observations do involve cliché expressions of respect and adoration between humans and nonhumans, the book makes the case that they problematize dominant expectations about the importance of feeling and care in politics. Indeed, as Tore Fougner argues, international studies has neglected nonhuman animals to such an extent that merely highlighting the many roles animals play in assemblages of international power constitutes a valuable first step toward considering their influence on and contributions to a variety of significant global political processes.[134] In this case, nonhumans materially support humanitarianism while also opening new forms of contestation by exposing the fragile, odd, and haphazard way in which relations of care emerge. In doing so, they raise important questions about how to reformulate humanitarianism and how to move beyond it. These questions, about the trajectory of a practice, the role of the human in humanitarianism's self-definition, and the anthropocentric predicates of global power, provide both new avenues for contesting violence and avenues for solidarity in a world that is always already more than human.

OVERVIEW OF THE WORK

This book explores these themes by examining three different clusters of nonhuman animals that labor in humanitarian operations: dogs, rats, and milk-producing nonhuman animals like caprines (goats) and bovines (cows). In each case, the capacities of these nonhumans transform practices of humanitarianism by revolutionizing logistics and supply, establishing new forms of security, creating nutrients, and, more often than not, addressing the suffering of both human and nonhuman others. What links these different nonhumans is not their genetic identity, nor their integration into human sociality, but the way they generate transformations in the conduct of humanitarian operations. In nonlinear dynamics, a singularity constitutes a turning point where a given system undergoes intensive change and opens the possibility for radically divergent, novel behavior. Classical chemical singularities are the boiling point in water or the temperature gradient that introduces turbulence (generating hurricanes) in otherwise placid equatorial oceans.[135] Singularities also emerge in social systems when the organization of political and economic life starts to behave differently and establish new patterns. They are the sign of a problem brewing, and they lead to the development of new concepts, novel questions, and emergent relations. The appearance of nonhuman animals in humanitarian operations constitutes this type of singularity. Nonhuman animals change both the content and form of humanitarian services in significant ways and enable humanitarian services to reach new, and assist, populations. Nonhuman humanitarians likewise stretch the ethical and political boundaries of humanitarian institutions and introduce questions about the distribution of care underlying problems addressed by humanitarian politics. Fundamentally, the involvement of these nonhuman animals generates a problem within the concepts, dynamics, and institutions of humanitarianism.[136] Beyond humanitarianism, they show a remarkable capacity, arguably breaking myths about the human origins of pacific behavior and political justice by revealing that these are multispecies outcomes.

These transformative effects are often subsumed by anthropocentric reason and anthropocentric feeling. At best, this reaction weakens the ability of humanitarians to consider how nonhuman animals contribute to the creation of better practices, and at worst, it means that humanitarians actively exploit the labor, suffering, and death of nonhuman animals. In some cases, nonhumans receive recognition for the gifts they provide human communities. In others, humanitarians expressly devalue nonhuman animal labor and justify killing and other forms of violence. However, if humanitarianism problematizes the human–nonhuman distinction and uses this insight to transform itself, then humanitarianism may join a host of other political efforts to challenge planet-wide mass extinction, ecological destruction, and entanglements of gratuitous violence that affect humans and nonhuman communities. It may also contribute to a dialogue about what ought to be left behind in the event of an extinction, not just in the sense of the life or remnants of a particular species, but the set of multispecies connections that might persist after this period without seeking to colonize these remnants or force them to cleave to images of the world defined by the human.[137]

Each chapter explores a specific practice of nonhuman humanitarianism, explaining how each group of nonhuman animals became participants in humanitarianism, while also trying to analyze the conditions of possibility for and consequences of this movement. It is a position that requires speculation because, in many cases, humanitarian organizations do not systematically analyze their relationship to nonhumans in this way, nor has the appearance of nonhuman animals, as a political force, been evaluated as a phenomenon occurring in multiple sectors of humanitarian services. As a result, the book places more emphasis on the ethical and political stakes of humanitarian involvement with nonhumans and less on the details of humanitarian policy. The text does not analyze specific empirical deployments of nonhumans in depth in favor of exploring how nonhumans became humanitarian and why nonhuman animals sometimes meet considerable barriers as participants in humanitarian services. Indeed, one of the immediate takeaways of this project is the remarkable divergence among humanitarian organizations when they are confronted with nonhuman difference. In some sections, this account is quite critical, illustrating how anthropocentrism tacitly or openly produces violent death for nonhuman animals. However, this reading should not be confused with a rejection or hostility to humanitarian efforts. Part of the ethos of agonistic respect and presumptive generosity discussed earlier is taking it as an article of faith that humanitarian efforts are also working toward a contestable version of the good and, on this basis, are capable of some novel forms of generosity. Hence the goal is not simply to critique but to reframe ongoing humanitarian practices as experiments in multispecies justice, partly to show how they involve ethically contestable and problematic relations to particular nonhuman animals; partly to demonstrate how the forms of recognition, care, and concern emerging from within these interactions may transform humanitarianism; and last, to show how they generate different possibilities for contesting anthropocentric power. By highlighting these processes occurring within humanitarianism, showing their only ambivalent relationship to humanitarian grammars, the book makes the case that there is something worth valuing in experiments in nonhuman humanitarianism. The remainder of the text traces the role of several nonhuman animals to show the distinctive capacities of nonhuman humanitarians impacting humanitarian politics.

The first chapter examines explosives and land mine clearance dogs. It describes the historical deployment of dogs in predatory warfare.[138] Land mine detection dogs developed as an offshoot of experiments during World War II before becoming a more widespread part of military, police, counterterrorism, and humanitarian operations over the ensuing decades. The chapter uses a biocultural lens to examine why dogs are excellent at the labor of explosive detection. Drawing on contemporary scientific research, dogs have multiple preadaptations to the work of demining that, when combined with behaviorist training regiments, dramatically improve the outcomes of many land mine clearance operations. Understood through the prism of anthropocentric reason, dogs are simply better detection tools than many mechanical devices, depending on the environment. However, the chapter argues that dogs open an entirely different sense of the mine field as a violent ecology, one where the potential for contingent death and injury emerges in ways that wickedly take lives even decades after the cessation of war. The chapter continues by arguing that dogs nevertheless introduce a form of joyful engagement with the horrific aspects of explosive ecologies. This joy creates a different set of sensations, a new relation to the significance of the mine field. Through a process of emotional contagion, demining dogs' capacity for joy amid a human-designed death trap rebounds into many different aspects of explosive detection operations in ways that exceed both the values of anthropocentric reason and the expectations of demining agencies. At the same time, this joy is far from unproblematic, because it has been cultivated deliberately and harnessed to make dogs more pliable in the face of mortal danger. The chapter concludes by wrestling with the ambiguous position of the demining dog as simultaneously an agent of radical politics, one that affectively reframes engagement with explosive territories, and a precarious subject of anthropocentric dominance.

The second chapter turns to the context of humanitarian rats. Like dogs, rats are now deminers. Bart Weetjens began the process of integrating rats into humanitarian operations in the late 1990s when he created Anti-Personnel Landmines Removal Product Development (APOPO), an organization that trains giant pouched rats to detect land mines.[139] Today, "HeroRATs" not only find mines but help with infectious disease identification and many other emergency situations. The chapter explores HeroRATs by juxtaposing the historical fear of rats as the enemies of humanitarianism, precursors to pestilence and disease, with their growing role in humanitarian operations. APOPO's deployment of rats not only requires scientific testing and determination but faces barriers to understanding rats as the agents of, rather than obstacles to, humane outcomes. This disgust, the chapter argues, emerges from principles deeply folded into agricultural practice that articulate strong aesthetic prejudices designed to enable the reproduction of identity. As a parasite, rats problematize these boundaries, raising metaphysical quandaries for agricultural societies, just as they eat foodstuffs. The chapter explores how the olfaction and sociality of rats enable new practices of humanitarianism because of the rats' sensuous capacities to interact with a zone or dimension of microfauna and bacteria otherwise imperceptible to humans. Understanding these capacities, the chapter argues, requires a shift away from an emphasis on agricultural politics and toward a vision of the world as contingent ecological entanglement. This section concludes by building on this insight to show how humanitarianism implicitly follows the logic of a legal order and, consequently, positions rats as giving gifts to human communities. These gifts point toward the gap or incommensurability between rat and human encounters, but, the chapter contends, paradoxically, this gap is crucial to articulating a model of multispecies ethics.

The third chapter examines goats, cows, and other milk-producing mammals. Popularized by Heifer International, several organizations now directly supply milk-making mammals as the physical substance of humanitarian aid. According to the organizational literature, caprines (goats) and bovines (cows) offer several benefits to communities or families in need, including providing consistent protein-intensive sustenance, maintaining sustainable farmland, and other economic benefits.[140] This chapter examines the introduction of dairy or farm animals into humanitarian services. It describes the historical rationale for supplying milk and analyzes how, unlike rats, goats and cows never factor into paradigms of gifting in this context of humanitarianism. Instead, nonhumans, despite doing the labor of producing milk, are curiously omitted from the economy of the gift themselves. The chapter argues that, in this case, anthropocentrism itself is one of the products offered by humanitarian organizations. Here, unlike the examples of dogs and rats, the goal of dairy farming is to eliminate a condition of subsistence by ensuring a consistent supply of protein (in milk and meat) and economic goods (in the exchange of animal products). Underneath fears of subsistence are broader metaphysical anxieties about contingency and precarity that are similarly inherited from agricultural politics. The irony the chapter exposes is that starvation is not only a contemporary political artifact resulting from inequity in food distribution in modern capitalism but a predicament resolved by the practice of instituting anthropocentrism. In this case, the delivery of nonhuman animals not only provides foodstuffs but mints new anthropocentric relations as a means of buffering human communities against contingency. This dynamic explains why goats and cows never appear as meaningful givers or agents of institutional change much as their labor dramatically reduces suffering.

The final chapter examines three implications of nonhuman humanitarians. First, it examines the limits of humanitarian principles considering this evidence and addresses the questions of how to address anthropocentric reason and extreme forms of violence. This relation-ship, the conclusion shows, is a condition of possibility for many other violences of humanism and humanitarianism. Second, it explores what it might mean to remake the politics of humanitarian care in relation to the affordances of nonhuman animals and the possibility of multispe-cies justice. Although it would be a mistake to romanticize nonhumans, there are key things that nonhumans disclose about the possibilities for care in a condition of weird violence and strange ecology. Finally, the book concludes by analyzing how nonhuman animals force a redefini-tion of humanitarian practice by loosening up the insistence on human life. In doing so, they raise the question, what would a different politics, emerging from but moving beyond humanitarianism, look like—one that more directly considered forms of life that do not, for various biological, aesthetic, social, cultural, or ideological reasons, resemble the human as it has been imagined? The chapter argues that current humanitarian relations are untenable by comparing the position of dogs, a companion in humanitarian labor, with rats, who give without companionship and are protected, and with goats or cows, who are consumed for anthropocentric reasons. What this illuminates is a double bind characterizing humanitarian involvement with nonhumans. This double bind intermittently exposes nonhumans to both extreme violence and saving grace. Both positions ultimately preclude a more substantive engagement with nonhuman humanitarians and humanitarianism's varying assumptions about the gift, law, and ethics. Instead, the chapter argues in favor of a more robust account of nonhuman communication that contests not just specific prac-tices of humanitarianism but humanitarian concepts. As such, the chapter concludes by arguing that the intensities of existing, entangled forms of care already show a model of generous politics beyond humanitarianism.

Humanitarianism is a comparatively recent historical invention, one that has undergone rapid transformation in its two plus centuries of self-stylized existence. This transformation has been buttressed by other political changes, including industrialization, colonization, logistics, digita-lization, and revolutions in armed conflict. Nonhuman animals constitute only a small part of the ensemble of humanitarian practices. At the same time, their role raises important questions about the most basic aspects of humanitarianism. Attending to the implications of nonhuman labor is vital to thinking about the future of humanitarianism, especially if emerging problems, such as global warming; racial, gendered, and class inequities; mass atrocities; and mass extinction, remain enduring features of global politics.

1

Dogs and the Politics of Detecting Explosives

Dia moves across a field of dirt through bright sunlight. In the distance, sparse grasses and trees line a recently cleared field. The space she is in is unremarkable, a patchwork of rubble potentially filled with metal or plastic debris, buried treasures, or unexploded ordnance. Dia moves fluidly through this place, pausing to examine objects of interest to her. She is largely unaided, if not unaccompanied, in her work. Her job is confined to this location today, but tomorrow, she will be somewhere else, facing a different terrain and a new group of challenges. Yet, Dia does not shy away from this work. Indeed, she was bred, born, and trained for it. When she finds what she is looking for, she communicates to her companion, and the crew begins the painstaking work of demarcating, extracting, and deactivating an explosive device. Dia is part of a much larger team of deminers, but she is currently the only land mine detection dog working in her party.[1]

Dia's job is to lead her team through spaces of the Cambodian countryside to locate mines and unexploded ordnance: the remnants of internecine and imperial warfare. These devices injure, maim, and kill decades after the cessation of formal hostilities between warring parties.[2] Dia's role is to guide the demining crew to the locations of this ordnance. She is particularly adept at this task, as the mines she finds were designed to avoid detection by human perception, covered with earth, overgrown with foliage, or scattered with so many fragments that it is almost impossible for a metal detector to find them swiftly. Moreover, Dia is quick, strong, and precise. She can pinpoint the smell of specific chemicals from a great distance and rapidly move to the site. She also appears to find joy in her work moving through this space, seeking and finding explosives, and assisting her demining group.

Dia was a real, living Belgian Malinois demining dog. She died from cancer after four years of service as part of the Cambodian Mine Action Centre (CMAC).[3] Despite her daily exposure to explosives, Dia, like every dog employed by CMAC and most demining dogs globally, was never mortally injured in her line of work. After her death, CMAC performed a funeral for Dia and buried her at the Landmine Museum in Siem Reap.[4] Land mine detection dogs like Dia often receive widespread recognition for the labor they perform through public dedication, memorials, and other affirmations.[5] This formal recognition indicates that detection dogs are understood as quasi-political entities and are valued for their part in creating safety for a community. In the past twenty-five years, dog detection crews in Cambodia alone have cleared hundreds of kilometers of minefields and reduced mine exposure for thousands.[6] Dia's labors vastly improved the speed and safety of these operations. In many respects, the acknowledgment of Dia's work reflects her tremendous value as an agent of humanitarianism. Yet even this acknowledgment is marked by anthropocentric reason and anthropocentric feeling.

This chapter examines the demining dog as a type of nonhuman humanitarian. It describes how dogs were introduced into explosive detection and outlines why this practice became more widespread. Dog explosive detection is now a global phenomenon, with hundreds of dogs employed by humanitarian organizations in many different countries. Dog detection is considered such a vital service that special contracts frequently protect these dogs, guaranteeing their health care and quality of life, in ways that are historically unprecedented for most nonhuman animals.[7] However, this chapter argues that humanitarian accounts of dog detection are framed by anthropocentric reason and that dogs are placed in these roles because they are understood as good instruments for explosives clearance. Thus, despite the dogs' celebrity status and formal recognition, humanitarian discourses often frame demining dogs reductively, as obedient laborers, but not as nonhuman animals that navigate minefields as complex political spaces. The chapter shows how this outlook obscures both the significance of the minefields as a political ecology and the ways that dogs not only improve but transform the horizon of humanitarianism. In particular, the chapter reveals how new forms of joy and companionship with dogs serve both to enable the possibility of demining and to interrupt its tragic framing.

THE CELEBRITY OF DOGS

Of all nonhuman animals, dogs have probably received the greatest amount of public attention alongside cats and exotic megafauna.[8] In many places, dogs are partners in the creation of human families and communities.[9] This role secures a place for them in public dialogue. They are, in Donna Haraway's terms, the quintessential "companion species."[10] But "man's best friend" only recently achieved this notoriety.[11] Even in Western contexts, within living memory, dogs were understood primarily as working animals that helped with farming and hunting.[12] However, as both natural and social scientists learn more about the historical connections between dogs and humans, simplistic narratives about dog–human relations become harder to maintain. For instance, both dogs and humans have been altered through a process of coevolution. This fact, now available in virtually every popular science text on dogs, suggests that dogs have not only played many social roles beyond herder and hound but influenced human development as well as the subsequent interaction between humans and other species and ecologies.[13] Social scientists have similarly challenged classical presentations of the dog–human relationship. For example, the hunting abilities of dogs have been used not only to chase birds or rabbits but also to hamstring soldiers on the battlefield, to extend the gratuitous violence of slave plantations, and to institute racial difference through the threat of violence.[14]

Dogs are also the subject of a variety of public debates about their role and activity in human societies. From concerns about dogfighting and animal welfare to the insidious construction of pit bulls as neighborhood dangers to city ordinances and signs legislating fecal matter, dogs are at the center of many political disputes at local, national, and global levels.[15] Discussions of dogs in public and global space reproduce disparate class, gender, and racial anxieties.[16] The publicly acknowledged consequences of dog–human cohabitation constitute only a sliver of the political involvement of dogs. Dogs also show up both literally and figuratively as voices of dissent and contestation. Consider the case of Loukanikos, or "Sausage," the Greek riot dog who repeatedly joined protestors to bark at police and parliament on the streets of Athens, as documented by countless videos.[17] Fictional dogs explain important lessons about politics, freedom, and equity. For example, the film *Milo and Otis* depicts dogs as obedient and dutiful, giving voice to the emotional tug of normative visions of home life, contrasted with the self-organized canine revolution of *White God,* the line of flight Tramp seeks in his desire to see "a great big hunk of world down there, with no fence around it" in *Lady and the Tramp,* or the resilience and anti-authoritarianism of Chief and Spots in *Isle of Dogs.* While these are anthropomorphic dogs, the mere fact that audiences understand the dog as the appropriate agent to deliver these messages indicates the degree to which dogs participate in politics.

In the United States, dogs have also received formal recognition for their labor in many human-dominated endeavors. From the use of bomb detection dogs and attack dogs in security contexts to service dogs to the growing role of dogs as emotional support animals, humans pay unprecedented attention to the labor of dogs.[18] Even President Donald Trump, the first American president in generations to break the norm of bringing a dog with him to the White House, made a spectacle of praising the labor of dogs in the raid against Abu Bakr al-Baghdadi.[19] More recently, Patron, a bomb-sniffing dog, has received global notoriety for his work countering explosives during the war in Ukraine.[20] Of course, this public awareness does not prevent dogs from being exposed to mass violence, as a glance at the numbers of dog abuse cases or the number of dogs abandoned to shelter life (and likely death) reveals.[21] As Sue Donaldson and Will Kymlicka note many times, "therapy and assistance animals are not trained to develop their own potential and interests, but moulded to serve human ends. . . . Animals with specific tractable temperaments are identified early, and pegged for future roles. Training, often very intensive over many months, involves significant restraints and confinement, and, frequently, severe correction and deprivation. Even so-called positive reinforcement is usually thinly disguised coercion."[22]

Moreover, the heightened concern for the welfare of dogs occurs against the backdrop of the unprecedented destruction of nonhuman life-forms, including multiple species that persist for the purpose of industrial meat production, often for the consumption of canine companions.[23] Companion species accent and extend human capacities in different ways, but they also more often than not reflect the liabilities and exclusions that dominate human societies. Anthropocentric feeling toward dogs, which encourages and idealizes dog–human kinship, finds a complement in anthropocentric reason that reduces other mammals to what Carol Adams so aptly called the "abstract referent" of meat.[24] As Cary Wolfe argues, proximate animal others are often easier to acknowledge and embrace, while more radical animal others, the sea slug or the tardigrade, are much harder to think about in terms of ethics or justice.[25] This observation is an important cautionary point because the arguments herein follow a somewhat well-scripted path that points to the singular benefits of human–dog collaboration in the context of humanitarianism. In making this case, it is worth remembering that the focus of humanitarian efforts generally occludes any investigation of the background conditions of nonhuman animal suffering and death that limn human–dog relationships.

The two more limited objectives of this chapter are, first, to show that dogs in humanitarian services are defined and dominated by anthropocentric reason and, second, to demonstrate how dogs transform these practices in important ways. The problem with many popular discussions of canine capacities is that they become easy gestures that fail to interrogate anthropocentric principles. The reason "look at what the dog can do" is a popular genre of animal commentary (e.g., Chaser, the border collie who knew a thousand words, or Blodgett, the cattle dog that performs extensive "barkour") is because dogs are so thoroughly enveloped by human aspirations, their very species life (with the notable exception of strays) informed, designed, and disciplined by the ambitions of anthropocentric reason over the past several centuries. One of the goals of this chapter is to retrieve demining dogs from this all-too-human predicament by highlighting the way that dogs not only do human things well but, in fact, change humans and humanitarian activities well beyond the confines of human perception and anthropocentric value. Exploring the minefield also shows how much presumably human action involves ecological complexity, where dogs can display more fecund, generous, and joyous relations than anthropocentric forms of humanitarianism. Dogs enter this milieu in ways that change not only the success of demining efforts but the broader ecological and ethical relations with martial spaces.

MAKING THE DOGS OF WAR

Dogs have long held specialized roles in military organizations alongside other nonhuman animals. Noted for their tracking abilities, dogs extend practices of what Gregoire Chamayou calls "cyngetic warfare" by making it possible to hunt an elusive adversary across inhospitable terrain using imperceptible marks.[26] Dogs' olfactory capacities extend armed conflict by contributing their perceptual field to military practice. This relationship emerged from the much longer coevolutionary links that humans and dogs formed in the process of tracking and hunting other nonhuman animals. However, in the context of armed conflict, dogs were trained to acquire more militaristic functions. Specifically, dogs were bred and conditioned to attack humans, to carry burdens across unfavorable environments (such as sled dogs), and to relay messages about terrain and human movement. Dogs remain a central part of several contemporary military practices and serve important symbolic and affective roles in armed conflict and martial discourse.[27] Dogs operate not only as a tool for tracking humans (a process now frequently outsourced to bioinformatics, forensics, and technological surveillance) but as a vector of speed and sensation that persists on the contemporary battlefield. In explosive detection work, this includes new training to identify improvised explosive devices and other modern technologies.

In the American context, training dogs to detect explosives emerged as an unintended consequence of research into discovering new roles for dogs during World War II. William A. Prestre, a Swiss expatriate, offered to train a small number of dogs and handlers to perform several new military functions, including attacking enemy soldiers.[28] The American military embraced Prestre's idea and opened a small space off the coast of Mississippi, ironically known as Cat Island, to explore these possibilities. At Cat Island, the speed, olfactory senses, and other abilities of dogs

Grady, an improvised explosive detection dog, waits for a command from his marine trainer during an Office of Naval Research technical demonstration. U.S. Navy photograph by John F. Williams.

were explored in detail. Researchers also examined whether dogs could be trained to detect Japanese soldiers based on odor, a study premised on the racist assumption that bioracial differences defined the human body and could facilitate targeted dog attacks.[29] In essence, the project sought to mobilize dogs as a new military force in racialized global conflict.[30] In doing so, Prestre was updating, but extending, the practice of using dogs to reify the boundaries of race as dogs had previously been deployed to consolidate racial formations in the context of slave plantations. In these contexts, the terror of being chased by dogs became a way of sealing off the possibility of flight from plantations and creating spectacles of racial terror and anti-Blackness.[31] Prestre's proposals were built on a similar presumption that dogs' olfactory abilities could solidify ontological boundaries filtered through biological understandings of racial difference. Explosive detection dogs grew as an offshoot of an attempt to design and perfect mechanisms of racialized warfare. However, these studies ultimately proved futile, because they were premised on illusory notions of biological and racial difference, but were swiftly replaced by efforts to train dogs to detect explosive materials and perform other sentry functions, as these were becoming an important part of the Allied campaign against Germany. This newer project, dubbed the M-Dog Project, successfully introduced dogs into the field in 1944. In these experiments, dogs were unable to reliably detect bombs. Only after further study, following the war, did researchers confirm that dogs could distinguish the chemical elements of explosives if subjected to the proper training regimen.[32]

During this period, dogs also acquired several other tactical and strategic roles in military life. Dogs were deployed to help deliver messages across unfavorable ecologies. They scouted territories and gave reports about enemy positions. Dogs also featured in combat on multiple occasions. One of the principal difficulties of introducing dogs to war was their lack of familiarity with an intense environment filled with gun and mortar fire. Consequently, dogs played largely peripheral roles in combat operations and were often confined to guard duties. Dogs acquired new positions in tropical armed conflict during the American island-hopping strategy in the Pacific.[33] There dogs could move much more swiftly and successfully through densely forested areas than their human counterparts. Following World War II, these preliminary training programs and the successful deployment of war dogs led to a vast expansion in the study and behavioral science of dogs.[34] Slowly, the lives of dogs became a more precise object of martial knowledge as canines integrated into and specialized as laborers in the armed forces. The inclusion of dogs in military life also enabled later deployments of dogs by police, security, and counterterrorism efforts.

The first appearance of dogs in humanitarian demining operations likewise developed as an outgrowth of military practice. Beginning in the 1970s, dogs started to accompany several humanitarian organizations into the field.[35] At a technical level, humanitarian agencies employ dogs in a way similar to how military institutions use dogs when they perform explosive detection. Each individual dog is typically accompanied by a single handler, who assists the dog in the field. This integration of dogs into humanitarian practice produced a series of studies on dogs' capacity to detect explosives. At present, dogs labor in explosive detection services as part of the United Nations Mine Action Service, the Marshall Legacy Institute, the U.S. Agency for International Development (USAID), Ronco, Norwegian People's Aid, and a host of other organizations. Many humanitarian organizations maintain separate breeding and training facilities for the dogs, investing in each generation of dogs to ensure that these canine services remain viable in the future. Detection dogs are frequently selected by organizations from these litters, which are carefully maintained to enhance tameness, sociability, trainability, and olfactory sense. Those dogs that show an aptitude for explosive detection receive ample living space, a proper diet, and medical attention, as well as suitable rehoming when they retire from their work as deminers.[36] Moreover, because of the dog's broader notoriety as a companion species, humanitarian demining organizations even use their dogs in advertisements as a method of attracting popular support.[37] Some humanitarian nongovernmental organizations, including CMAC, have offered visitors to their facilities "meet-and-greet" sessions with their demining dogs as a method of

An explosive detection dog and handler with Norwegian People's Aid in Bosnia and Herzegovina. Courtesy of Norwegian People's Aid.

procuring donations to the organization. In this way, dogs work not only as deminers in the field but as attractors that showcase the valuable work performed by demining organizations. Interestingly, the introduction of demining dogs into this role also led humanitarian organizations to publicize how they treat dogs, documenting the breeding, training, and caretaking processes. Mine clearance organizations thus appropriate and manage the entire life cycle of dogs as a resource: for clearing explosives, for advertising, and for regenerating these capacities.

This brief overview of the military and humanitarian deployment of demining dogs raises two important questions about the political status of demining operations. First, does the mutual investment of humanitarian and military organizations in the explosive detection abilities of dogs reveal a greater proximity or connection between these two, ostensibly opposed, forms of practice? Here the movement of dogs from military to humanitarian contexts shows the ways in which, as Eyal Weizman argues, humanitarian efforts are conditioned by military practice.[38] In Weizman's examples, the logic of lesser evil and the structure of humanitarian outreach draw from and reproduce militaristic distinctions to define their roles. In effect, humanitarian outreach, while seeming compassionate, not only accommodates but often buttresses militaristic practices. In the context of explosives clearance, this process is apparent as humanitarian services build on but also contribute to knowledge production that may be applied in a martial environment. This implicit exchange of knowledge is not intrinsically problematic. Yet, the connection between dog detection, military affairs, and securitization rarely receives public commentary or contestation. Indeed, humanitarian discourse treats dog detection as a distinct, celebratory process of saving lives (akin to search-and-rescue dogs), all the while downplaying how this capacity is a repurposing of a military function and, crucially, that these same capacities are deployed in landscapes defined by the detritus of war. Put differently, humanitarian demining efforts intervene on a temporally and spatially complex zone of armed conflict. When they do so, they employ knowledge, such as dog training procedures, that also emerges from martial practices. Implicitly, humanitarian organizations use the means of military violence to remedy the remnants and debris of imperial warfare.[39] However, the deployment of demining dogs in advertising discursively separates humanitarianism from a more substantial, antimilitarist politics by treating humanitarianism as the mission of saving lives without directly connecting this practice to legacies of warfare. The point is not that humanitarian operations are hypocritical in their self-presentation. Rather, explosive detection dogs reveal how deeply implicated humanitarian practices are with the expansive militarization of social and ecological life as well as broader regimes of colonial and racialized violence.

The second question this relationship poses is how these humanitarian and military organizations understand dogs when they breed, train, and deploy them for demining operations. In these actions, each organization captures and cultivates the biosocial capacities of dogs for the express purpose of saving and extending human life, but each organization is also engaging in a calculus that exposes dogs to injury, debility, and death. Ultimately, despite their notoriety, the lives of detection dogs are directly subordinated to human ends. This process rests on an affirmation of demining dogs to advertise the virtuous character of humanitarian services while, at the same time, devaluing the lives of demining dogs in the process of harnessing their distinctive olfactory capacities to govern threats to human communities. These two gestures exist in a complementary relationship with one another. Absent the open celebration of demining dogs, the crude anthropocentrism of demining operations may be contested because of the close, companionable, and even intimate relationship between dogs and some human constituencies. Anthropocentric feeling constitutes the paradoxical vector through which dogs become legitimately exposed to threats of explosion. In addition, humanitarian organizations capitalize on the presence of dogs in demining, and their exposure to danger, to help humanize humanitarian projects. They do so by doubling down on anthropocentric feeling, deploying the sympathy created by dog–human relations to produce empathy for demining projects. Because a large constituency, disproportionately situated in the Global North, forms emotional connections through images of dogs, detection dogs offer a conduit to establish sympathetic links to the problem of explosives. Dogs become the means of humanizing these efforts, but this very humanization is what, in turn, sends demining dogs into fields of danger.

Unlike other nonhuman animals, explored later in this book, that struggle to be considered partners in humanitarianism, the case of demining dogs shows how the humanization of something deemed nonhuman does not necessarily lead to protection but instead may jeopardize nonhuman life. Humanization becomes a mechanism producing multiple forms of compassion and identification, but with new risks for dogs.

DOGS, BOMBS, AND SENSATION

Given the deployment of explosive detection dogs by military, security, and humanitarian organizations, dogs must have excellent demining skills. But why? What makes a dog successful at detecting explosives? Since they became active participants in contemporary armed conflict, many studies have demonstrated that dogs have several capacities that make them particularly good at finding bombs.[40] The most widely discussed are dogs' olfactory capacities. Although everyone knows that dogs smell better than humans, the degree of difference is hard to contemplate. A dog's olfactory glands tend to be several orders of magnitude more sensitive than a typical human's. To illustrate this discrepancy, popular science publications sometimes point out that an average dog can smell a single teaspoon of sugar dissolved into a liquid solution the size of an Olympic swimming pool. Several aspects of dog olfaction enable this degree of sensitivity.[41] Dogs smell in three spatial dimensions and can detect intervals, changes in time and movement, based on smell intensity.[42] Technically, humans also smell in these dimensions, but not with nearly the same depth as dogs. In addition, dogs' nasal cavities are augmented by a vomeronasal organ, an olfactory membrane that enables dogs to detect scents more easily and that became vestigial in humans at an earlier point in their evolution. Finally, dogs breathe differently. A dog sniff, unlike a human breath, facilitates a continuous circulation of air within the dog's nose, allowing it not only to detect the presence of the smell but to give the odor a spatial contour and trajectory. This respiratory process allows dogs to track the motion of other mammals, birds, and reptiles but also likely helps identify explosives, which emit smells but remain otherwise inert. These features of dog olfaction help explain why dogs introduce new possibilities to demining work.[43]

Dogs are effective at interacting with explosives and land mines because they can scent the aerosolized residues of trinitrotoluene (TNT), dinitrotoluene (DNT), plastic explosives (C-4), and other elements of explosive ordnance. The components of an explosive like TNT or DNT dissolve at variables rates as they interact with wind, rain, soil changes, and other life-forms where it is embedded. Exposure leads to atmospheric dispersion that allows dogs to pick up on the presence of explosives.[44] Dogs detect these chemical signatures because of the sensitivity of their trigeminal nerve.[45] This nerve allows dogs to register and recall smells to a degree far beyond human capabilities. With training, many dogs can detect TNT, DNT, or the elements of plastic explosives, but some excel at the task, depending on the density and dispersion of these elements in the environment.

Framed in slightly different terms, dog olfaction has a greater potential degree of power than human olfaction, but activating this potential in relation to objects in the built environment that are of special interest to humans requires significant practice.[46] Dogs need to be not only introduced and acclimated to the distinction between odors but also experienced at navigating environmental factors and avoiding the triggers for explosive mechanisms. This training requires dogs to learn how to ignore specific environmental stimuli to concentrate on finding explosives. Moreover, because of existing research on the relative olfactory capacities of different dog breeds, only specific dogs ever join demining teams. More generally, explosive detection training builds on a much broader literature on dog behavior and ethology, which coordinates programs of socialization and communication. It is worth noting that in some environments, dogs are not ideal deminers because of their reliance on smell. Temperature gradients, atmospheric pressure differentials, wind trajectories, humidity, levels of rainfall, and the presence of other human and nonhuman animals all can interfere with the quality of the detection process.

Nevertheless, dogs introduce a new sensory regime into the process of explosive detection. Land mines and other disguised explosives work as technologies of violence because of how they take advantage of certain normalized habits of perception and intelligibility. As Antoine Bousquet argues, for humans, the ecology of war typically advances through lines of visuality and thresholds of detectability.[47] A land mine or unexploded ordnance poses a danger because of the reliance of most humans (military forces and the everyday passerby) on the visual spectrum to ambulate within a given environment. The explosive turns this dependency into a liability and liability into debility and death, making the environment appear danger-free from the vantage point of the visual spectrum and exploiting the limitations constitutive of a structure of sense perception.[48] Dogs, in contrast, rely on smell as a much greater part of their sensorimotor processing and locomotion and move more safely and with greater exactness in the same space.

Although dogs are famous for their olfaction and rely heavily on this sense in detection operations, olfaction is not the only capacity that dogs use in completing these tasks. Unlike most bipedal humans, dogs use quadrupedal ambulation, which gives them the ability to cover much larger areas on foot. As a generally smaller mammal, dogs are more successful at fluid movement through nonurbanized ecologies. Dogs develop a different center of gravity and possess better balance than human beings on some surfaces, which makes ascents and descents easier to traverse. Though dogs do not form the same kinds of visual intensities as humans, they are excellent at identifying human modifications to an environment. This ability often enables dogs to identify trip wires or other forms of camouflage that would be harder to sense for their human counterparts. In addition, although dogs do not complete the engineering work of dismantling a land mine or explosive, they have occasionally aided clearance operations by filling in the cavities formed in demining.[49] These abilities enhance the speed and reduce the environmental impact of demining operations as dogs move with greater efficiency than human deminers.

It is important to interpret the olfactory senses and other capacities of dogs as an assemblage that emerged from specific historical conditions of possibility forged through dynamic interaction between dogs and their environment.[50] In evolutionary-developmental terms, the capacities of dogs were crafted from a series of pressures for hunting, scavenging, and resistance to disease gradients as well as parasitic, commensal, and symbiotic relationships. Since their coevolution with humans began, specific features of dogs' sensory perception and modes of sociality have been further developed through cohabitation and, more recently, intensive breeding regimens and the formalization of behavioral science.[51] While describing each of the ecological and social pressures that informed canine evolution and dog training is beyond the scope of this chapter, the point is that the dog's sensorimotor apparatus is an emergent, relational product with unique affordances and aversions to different ecologies. Dogs and their capacities are distinct but fluid and capable of change, partially closed because of the hard pressures of deep geological time and partially open to rewriting and therefore to forming new relations.

Dogs actualize what Manuel DeLanda calls the "blueprint" of animalia, involving many isomorphic features, including ambulation, a brain capable of complex problem solving, sensitivity on the dermis, and so on, but they also differentiated from other nonhuman animals as the species evolved over time.[52] In this regard, the capacities of the dog are not so much special as distinctive, emerging from a singular historical lineage, developed from mutation and phylogenic differentiation. In a sense, the "capacity of their capacities" to enter new relations in combination with the rise of human life (and much later contemporary development) reveals that dogs actualize capacities as part of an open assemblage. While human perception typically looks at these capacities as tools that can be favorably deployed to human ends, a dog's capacities also reveal potentialities that complicate and expand the horizon of humanitarianism. In addition, while dog capacities may be treated as technologies, they are also rooted in a form of sense experience. Though, as Thomas Nagel famously argues, this experience is incommensurable to human perception, dogs experience their capacity to act in relation to surrounding ecologies in ways that incite a variety of emotional responses.[53] The environment itself is a series of differential intensities that trigger these affective changes in dogs. As I discuss later, these affects are a crucial part of why dogs can successfully navigate minefields and other spaces of distended armed conflict.

In relying on the abilities of dogs, humanitarian operations harness dogs' aesthetic and affective apprehension of the world, or what Jakob von Uexküll famously called their "*Umwelt.*"[54] The dog's ability to smell or move is not just an instrument for better locating an explosive, although this is the function for which dogs are deployed in the field; rather, this sensory regime builds from an entirely different composition of reality, providing dogs' sensual production of time, space, objects, events, and politics. When humanitarian or military organizations integrate dogs into demining practices, they not only substitute one tool for another but draw on an entire process of world building to apprehend the existence of the explosive in terms. Put differently, the dog's world becomes a set of perceptual potentials from which to extract the possibility of explosive detection. However, the deployment of dogs into the minefield does far more than simply enhance demining operations' power to locate. To understand these effects, it is necessary to explore a bit more thoroughly what constitutes a minefield as an assemblage where dogs and humans traverse, see, smell, perish, and live.

THE WEIRDNESS OF EXPLOSIVE ECOLOGY

Most of the research on dog detection seeks to analyze and exploit the "natural abilities" of dogs. However, describing something as natural is an effort to assign an unchangeable essence or identity to a thing. If the dog is a "natural deminer," it is because it is supposed to have a fixed essence or set of attributes that predetermine its capacity to detect explosives. Because political agency is not one of these attributes, dogs can be leashed and muzzled by humans for their own purposes without significantly harming the dogs or without jeopardizing a more meaningful, more valuable form of life. In this way, discourses on dog ontology and on the natural capacities of nonhuman animals normalize control over their canine breeding, upbringing, and exposure to danger. However, paradoxically, the need for breeding, training, and cultivation of both sociality and detection skills indicates that these are not natural propensities but potentialities that emerge through historically contingent interactions defined by force. Moreover, the claim that the dog is a "natural deminer" also implies the existence of something like a "natural minefield." As such, in this discourse, explosives are implicitly not seen as part of the detritus of colonial or imperial conflict deliberately enacted and (in most cases) left to fester and wound generations after the formal cessation of hostilities but rather as a given environment, a set of taken-for-granted conditions that simply need to be dealt with by "natural deminers."[55] The dirty machinations of colonial violence disappear into doublespeak about the laboring propensities of dogs. As such, in the process of naturalizing dogs' capacity to detect explosives, these discussions also normalize a postcolonial predicament defined by sporadic, unpredictable, and distributed violence.

It is worth unpacking these concepts further. The philosopher Graham Harman argues that "nature is never natural and can never be naturalized."[56] The very idea of nature, and therefore natural capacities, is an anthropomorphic fantasy that views entities as harmoniously coexisting from the vantage point of human experience.[57] This fantasy commits what Alfred North Whitehead terms the "fallacy of misplaced concreteness."[58] Instead, "nature" is a series of ecological relations, partial aesthetic interactions characterized by ontological indeterminacy. By this, Tim Morton and others argue that distinction between appearance and being, or what a thing looks like and what it is, is ever indeterminate. Ecologies are weird because they involve an odd interpopulating of objects, minerals, plants, and animals that are all only incompletely disclosed and vicariously intermingling with one another.[59] There is no consistency in an ecology but a group of disjointed catastrophes, phases of heterogeneous change, and strange couplings between different beings. Patterns emerge in the interaction between objects, but these interactions are always provisional rather than an inherently recurring process. Understanding dogs and explosives as having any kind of "natural" connection cheats on the complexity of ecological life. It normalizes the risks for dogs in the process of land mine detection and overlooks the fact that what might be called "explosive ecologies" are politically produced, ontologically indeterminate zones of armed conflict. The interaction of dogs with explosive ecologies reveals what might be called "weird violence" based on the distribution of precarity, uncertainty, and danger cooperatively produced by many actors in these environments.[60] Indeed, there is nothing natural about an explosive ecology at all. It is a built environment, but one that mirrors or mimics the appearance of an undesigned terrain. In the construction of this ecology, a calculation occurs about the persistent exposure of some lives to death and debility.[61] It is a sliver of what Jairus Grove calls a "savage ecology" or a form of politics "radically antagonistic to survival as a collective rather than discriminatory goal."[62]

Yet, the designed character of the explosive ecology does not deny the fact that it is a mobile assemblage. Explosive ecology, as a space showing the markings of deliberate human intervention, is an environment populated by all kinds of different machines, objects, and entities that together generate new epistemological uncertainties. These epistemological uncertainties are a part of their lethal capacity and subsequent danger of the explosive ecology. At the same time, land mines and explosives are also actants in their own right. As Leah Zani contends, explosives, land mines, and bombs should be understood as desiring agencies that seek to explode and erase the evidence of their own existence.[63] In this reading, the "priming" of a bomb involves a persistent, if constrained, will to detonate. A state, military organization, or individual that lays an explosive cannot know for certain what will trigger the explosion. The organization's use of the explosive depends on an epistemological move to produce spaces of danger and impassability structured by a combination of epistemological certainty, that the explosives are set in each place at a given time, and uncertainty, that they will contingently interact with other occupants of this space. Nonetheless, the organizations setting the explosive cannot reliably predict with certainty the actualization of any particular death or injury. Setting or dropping explosives thus presupposes a disregard for any singular loss of life and a general disdain for the possibility of life flourishing in proximity to the explosives, a kind of erasure of a necessarily unknown otherness based on broadly adversarial and historically racial-colonial terms that arguably constitutes the ecology's first act of violence.

However, the ontology of explosive ecology is even stranger than this. Any explosive involves layers of contingency, including the possibility of the explosive's self-activation. For instance, even the most deliberately set land mine depends on camouflage to be successful (unless the minefield exists as a signaling barrier outside of a military post, for instance). Camouflage involves reproducing the appearance of the landscape, but in the act of reproduction, a kind of enacted repetition occurs that invites contingency into the deployment of violence. The elusiveness, hiddenness, and randomness of this process are actually crucial to producing affects of terror and fear that make land mines effective mechanisms of control over human communities. The affects that emerge from explosive ecologies transform territories into danger zones and, by doing so, regulate and govern the possibility of movement. These effects make the deployment of mines a useful part of military strategy because explosives not only injure an individual person or group but create emotional changes that render specific lines of movement impossible as they engender environments of persistent terror. Ecological weirdness—is this a minefield or a "typical" forest? is this footfall firm because of earth or explosive?—is critical to the production of the land mine's ability to control and kill. These features also open explosives to drift, rot, growth, or erosion, all processes that may dampen or enhance the likelihood of an individual explosive detonating. The violence of explosives originates in some form in human decisions, but it is enacted by, performed by, and dependent on an "ontological indeterminacy" that subtends epistemological uncertainty.[64] Both the human-designed and environmentally enfolded aspects of explosive ecologies are worth exploring in a bit more detail because they illuminate some of the limitations of humanitarian practice and the unique potentials dogs introduce into processes of demining.

Human actions play a defining role in the making of explosive ecologies. Discarded or unexploded munitions are placed during armed conflict. Explosive detection and clearance is an after-the-fact response to the ruins of war.[65] Unfortunately, this framework for describing demining both overmines and undermines the importance of human involvement in mining. To unpack this point, consider the relationship between mining and warfare. In modern conceptions of armed conflict, war has been understood as an activity conducted between organized, professional armies linked to the state. Conflict takes place between these organizations within a defined (if disputed) set of temporal and spatial boundaries. Critical studies of warfare have demonstrated multiple limitations to this model of armed conflict. First, this conception of warfare is linked to one particular moment when modern states sought to secure a monopoly on legitimate violence.[66] This coincided with colonial occupation and, as an understanding of armed conflict, reproduces Eurocentric norms about the legitimacy and norms of war.[67] Second, though related, practices of martial violence, such as bombing and mining, are political and colonial acts.[68] In many contexts, imperial powers deployed explosives to generate precarity in formerly or actively colonized communities, using their indiscriminate violence to produce prolonged states of precarity and powerlessness.[69] Finally, war does not have stable temporal or spatial boundaries, nor does armed conflict necessarily follow from human designs. While the explosives and munitions that persist after the end of formal hostilities are not a part of "combat" or "battle" in the classic sense, they nonetheless involve lethal possibilities. An explosive ecology is often a paradoxical zone of pure combat in the absence of a battle or, conversely, a lethal battle space in the absence of combat. Either way, armed conflict becomes distended and participates in what Rob Nixon terms "slow violence" alongside the more rapid, spectacular forms of violence with which modern warfare is more often associated.[70]

If war produces its own temporal horizons, then human decisions cannot be held accountable for all events of violence according to strictest legal standards of intent. Indeed, it would be difficult to prove that a particular general, politician, or soldier authoring an act of violence, such as laying a mine, could know, thirty years later, that another individual person, one born after the conflict, for example, would be injured by the bomb.[71] In this sense, emphasizing human decisions in the creation of the explosive ecologies overmines, to use Graham Harman's language, human intentions in order to establish causal and moral lines of responsibility for an act of violence. This approach subordinates ecological weirdness to human agency and, by doing so, deprives explosives of an autonomy and potential for interaction that is a crucial part of their danger. At the same time, if war does not have discrete beginnings or ends, if it transforms the environment, and if it was initiated because of a priori determinations about the value of forms of life, then human decisions also play a crucial role in the commission of violence. Downplaying or dismissing the role of human agents would, conversely, undermine their importance in producing any case of violence. Intentions are crucial to the situation, but their power derives from a broader distribution of agency because of the assemblage of the explosive ecology, and moreover, it is further complicated by the ontologically contingent, partial interactions explosives have with the other people, nonhuman animals, plants, and objects in this ecology. As Jane Bennett explains, "human intentionality can emerge as agentic only by way of such a distribution. The agency of assemblages is not the strong, autonomous [. . . but] more porous, tenuous and thus indirect kind."[72] Consequently, human designs are key for the emergence of a violent event, but their causal involvement is vicarious, produced through the actions and entanglement of other objects and systems.[73] In this sense, the rhetoric of demining advocates in a reductive proposition when it makes the case that particular countries or leaders are evil on account of their use of explosives while, at the same time, arguments that attempt to limit the accountability of the states that set explosives are also incorrect when they push back against this narrative.[74] Both propositions paraphrase the vicarious, distributed character of this form of violence. These propositions also exploit, in very different ways, the relationship between ecological weirdness and the violence of these explosives to enhance or contest their accountability for these conditions.[75] In the end, specific human intentions constitute one of the conditions of possibility for the emergence of explosive ecologies, but these intentions are about enacting not just one act of violence but the production of a discriminatory future in which violence becomes environmental, and the environment agentic, in the commission of violence. Explosive ecologies thus engage in a preemptive violence on the future possibility for life.[76]

In this sense, explosive ecologies also exceed human intentions and have a strange way of reproducing themselves. For example, mines occasionally reveal a capacity to break with their conditions and act in new, unforeseen ways. Put differently, explosives operate through ecological means. First, the capacity for an explosive, such as a land mine, to injure depends on ecological aesthetics. Camouflage, the process that makes mines useful from a tactical or strategic perspective, depends on what Bousquet terms the "martial gaze."[77] Land mines function by manipulating the visual field to make the explosive undetectable and deadly. To do so, they mimic expectations about what a given ecology should look like according to normative models of human perception. Military strategy builds on structures of perception and habits of embodiment and ambulation. Second, the explosive capacities of land mines are a modification of existing ecological dynamics from the exothermic potential of nitrates to the pressure-based levers that trigger a device. They hinge on the uncertainty between appearance and being to trick a passerby into assuming that a dangerous space is safe or to render ordinary spaces so hypothetically threatening that they become unpassable. In this sense, there is no possibility of separating the land mine from an ecological milieu. Third, at the same time, explosives are also separate, observable things and not merely bundles of preexisting social relations. They may exist as present in each environment because of relational connections to humans, chemistry, manufacturing, and so on, but explosives can also participate in novel interactions that change their agency or degrees of power. Land mines, for instance, are sensuous entities, responding to the pressure of the foot, paw, or hoof. Moreover, a land mine may act through "malfunction" despite perfect machining. The triggering mechanism may rust in a sopping wet ecology. It may autodetonate due to various components enhancing the pressure on the device. Though analytically tracking these changes to shifts in particular properties of the explosive is useful, it is an ontologically dubious exercise, because explosive forensics often encounter problems that have no mechanical explanation in the engineering literature.[78] The field of engineering invented the term *accident* to describe an event that resists analytical explication.[79] By exploding, rusting, or deteriorating, the mine evinces potentials to betray the intention of its makers and even the properties that theoretically define it as a mine. A land mine that will not explode despite all the appropriate chemical priming is still a land mine, just one that thwarts human designs.

If an explosive is both subject to a strange network of shifting relations and a semiautonomous entity, then it becomes exceedingly difficult to point to the causal force in an ecology that is responsible for triggering an explosion. In a legal paradigm built on the model of the rational subject capable of self-authored, moral, and causal decisions, this poses an enormous problem, because it makes any proposition regarding strict morality or efficient causality dubious. All accounts are poor paraphrases of what is happening.[80] However, the weirdness is also the point. The violence of explosives is disguised and twisted, turning on the potential for ecology to hide causality and making the apparent absence of causality into the source of mortal danger. In fact, this point makes intuitive sense because it follows from the mine's capacity to produce deadly violence through camouflage. Indeed, it reverses the argument. In contrast to human engineers exclusively and intentionally creating camouflage, land mines reveal a potential to both relate and not relate to other entities; to kill by entwining appearance, disappearance, and interaction with a larger ecology; and to autonomously explode in ways that also break with these networks and connections. Militaries, states, and other violent actors exploit the limits of human sense perception when it encounters this thicket of phenomenological and ontological indeterminacy. In short, mines are deadly *technē*, a different mode of revealing the weirdness of ecology that may or may not disclose itself at any point. Weird violence is not a reason to excuse the violence of setting mines but highlights the need for a more complex account of the forms of agency involved in creating the conditions of violence.

In addition, the ecological weirdness of an explosive produces another important quality: its horror. The land mine's capacities to play with appearance, to elude human sense perception, and to make the unknown into the source of doom are uniquely troubling. On one hand, the mine interrupts what Adrianna Cavarero calls the "figural unity" of the human (and, although Cavarero does not go into this, nonhuman animals as well) by explosively dismembering the victim and, in doing so, challenging the presumptive integrity of a body.[81] On the other hand, the ecological weirdness entangles the victims of violence, keeping them in a zone from which they cannot flee because the very possibility of careless flight might incite an explosion.[82] The land mine is a trap defined by ecological webbing but also capable of contingent self-initiation. To enter an explosive ecology is to enter an arena where the environment may or may not, could or could not, destroy the body. Once this zone becomes a paradoxical space of known uncertainty, both movement within and exit from pose lethal dangers. However, unlike the horrorism that Cavarero envisions, where violence enters the public sphere, explosive ecologies involve a "wild public" in a more expansive, unstructured sense.[83] Communities afflicted by explosives live with an affective terror that may or may not always be felt but cannot be fled from, because flight opens the possibility of encountering explosives that dismember or destroy. If flight is impossible, then exit, in the form of wealth or relocation, is the only alternative to the work of land mine clearance. In the meantime, these zones of horror operate as a form of what Achilles Mbembe terms "necropolitics," constitutive of the capacity to place explosives and foster self-organizing dynamics of annihilation. Necropolitics consigns human communities to a position of ambient dread that is psychologically toxic and replete with lethal spontaneity.[84]

This section began by exploring the problematic proposition that a dog is a natural deminer. At the conclusion, it is worth revising this statement: dogs are not natural deminers; dogs are strange ecological deminers. They enter the same thicket of ecological relations and human-augmented spaces of horror, but they confront these ecologies differently than humans, without anthropocentric perceptions and affects. Instead, dogs often approach explosive ecologies with abundant, frequently observed forms of joy. This joy plays a key role in contesting both the overt violence of colonial armed conflict and the savage uncertainty of living in proximity to a deadly explosion.

HUMANITARIAN TRAGEDY AND THE JOY OF DOGS

Dog explosive detection is not natural but ecological, partial, and contingent. A dog's movement through a field, accompanied by human companions, is an exploration of the thick heterogeneity of sensations that constitute an explosive ecology. While existing scientific studies argue that olfactory sensation explains the success of dog detection, this perspective neglects the other capacities that the dog reveals in its encounters with explosive ecologies. Crucially, dogs often seem to find the work of discovering explosives joyful, fun, or playful. Many accounts of detection describe the importance of affective bonds in the production of demining work, including anecdotal narratives to the strategic calibration of behavioral analysis and training.[85] As Vicki Hearne contends, focusing on the possibility of happiness, joy, and play is important in the context of human–nonhuman encounters because "what interests some people is not the joy and intelligence and difficulty and difference of animals, but only their pain."[86] In the humanitarian context, operations are often assessed in terms of explosives cleared and lives saved—vital metrics, but standards that do not observe the qualitative influence of dog detection units. Of course, claims about the joy of dogs are mediated by human access and interpretation, but, if Hearne is correct, then the consistency with which humans observe affective changes in laboring with detection dogs suggests that some emotive difference materializes when dogs and humans cooperatively interact with explosive ecologies. Happy affects are the targets of behavioral regimes designed to instruct dogs how to engage a minefield through different schedules of reinforcement. Much like demining abilities, joy is not "natural" but produced, caught, and reinforced using incentive structures that help turn detection into a game of hide, seek, and reward.[87] However, priming a dog to appreciate demining is only one aspect of dogs' play and one understanding of what is ongoing in joy. The philosopher Baruch Spinoza opined that joy emerges as a body experiences affirmative relations with other bodies, whereas sadness develops with negative encounters with bodies that restrict one another's capacities.[88] Read in light of Spinoza's propositions, the dog's reaction is not reducible to an incentive program or to felt articulations of canine happiness but involves a set of pluripotent encounters between the dog and the other bodies in this ecology. Tracing each aspect of these encounters is impossible, but partial interaction with a field composed of multiple olfactory sensations likely intensifies a dog's sense of its own power, producing something like Spinoza's joy. Even within a strictly behavioral framework, these background factors constitute "setting events" that play an agential role in the emergence of a shift in the dog's behavior and disposition.[89]

Furthermore, in Spinoza's terms, the capacity for joy in the encounter with the explosive ecology creates the possibility of dog detection. If dogs found explosive detection labor aversive, intolerable, or sad, in Spinoza's sense, as many humans obviously do, then there would be far greater resistance to engaging in this endeavor. Instead, the dog's mode of access, in conjunction with human companions, whatever that involves, opens the explosive ecology as a site where joy becomes a felt possibility. The training regimes championed by humanitarian organizations contribute to this process by reinforcing specific actions and behaviors, but these reinforcement regimes also function by augmenting emergent tendencies for intensity, pleasure, and play. Training, much like ethical self-cultivation, involves the selection, production, and aesthetics of intensive experiences. In this case, the possibility of a dog's joy is both a driver and the product of the detection process.[90] Joy also functions as the means to "capture" the capacities of the dog because it is the possibility of joy that enables demining operations to deploy dogs successfully in the field. Joy thus operates as an implicit presupposition of the entire process of demining, working both as its condition of possibility and as a force that propels dogs into explosive ecologies.

In addition, the joy of dogs injects a different ecological disposition into the practice of land mine detection.[91] In a certain sense, the problem of dog explosive detection involves a classical political dilemma: "why do people fight for their servitude as stubbornly as though it were their salvation," or, in this case, why do dogs persist in exposing themselves to harm when they could resist this encounter?[92] For dogs, as for humans, demining is a possibly deadly activity. Explosive ecologies are spaces of contingent death and ambulant sadness. For many humans, they possess an aura of palpable horror, haunting trauma, and lingering wounds that saturates humanitarian literatures.[93] Dogs, in contrast, do not seem to find this work saddening. To be clear, this is not because dogs are necessarily ignorant of the consequences of explosion because they are also trained to avoid and attuned to triggering explosives when they are found. Rather, dogs relate to explosive ecologies in entirely different ways than their human handlers. While it would be impossible to describe this experience from a dog's perspective, the dog's involvement with humans produces affective changes in the explosive ecology because its engagement creates secondary effects that are directly observable and communicable for humans. Here a form of affective contagion, in which given emotional reactions radiate through an assemblage, moves from dogs to humans, changing the relationship to the explosive ecology in which both humans and nonhumans operate.[94] For instance, joy changes the disposition of human deminers as they labor in proximity to the horror of explosives. These emotive changes are an example of what Vinciane Despret calls an "embodied experiment [that] builds affinities, partial affinities."[95] The involvement of detection dogs interrupts the looming sense of tragedy, sadness, and horror. This interruption opens the possibility for human deminers also to experience a shift in disposition, converting their engagement with explosive ecologies from one predominantly defined by danger into a space of multiplicity that certainly contains danger but also discovery, games, comradery—the possibility of life unburdened by ambient doom.

In this sense, the dog's joy constitutes a means of communicating a new version of the explosive ecology as a political reality. Demining, rather than dangerous labor, is implicitly contested and therefore becomes a practice akin to a form of play or experimentation, a movement that involves a joint exploration of sensation rather than a confrontation with mortal danger. While a stern, moralistic attitude might decry these attitudes as inconsiderate in the context of an explosive ecology, these are critical dispositions for demining because they enable operations to work in ways that would be otherwise impossible. Where anthropocentric reason views the dog as a useful tool, ecological weirdness involves connections between dogs and humans that reshape the affective relations to all of the strange, partially disclosed entities that make up the explosive ecology. Speed and precision, which can certainly be institutionally captured and expanded, are by-products of this underlying, apparently palpable canine excitement.

While traditional accounts of emotion and affect view them as private affairs, affects constitute bodies, producing new rhythms even among interspecies encounters.[96] Dogs and humans are potentially primed through a variety of evolutionary, social, and cultural mechanisms to respond to the affective couplings that actualize through cohabitation and intensify through training regimes.[97] Put simply, dogs may change human affective dispositions, and vice versa.[98] Joy in an explosive ecology is not solely a property of canines. At a simple level, the sensation of the joy of dogs changes the perceptions of many human deminers. At a more complex level, the entire dynamic—movements, pathways, conversations, tone, and intensity—of land mine work transforms because of the assembly of dog–human demining teams. Humans do not mimic dogs when they experience greater joy in demining but resonate with a different atmosphere by joining dogs in this labor.[99]

This cooperative interaction challenges part of the necropolitics of the explosive ecology. In an assemblage of dogs and humans, demining crews expand their affective engagement with the minefield. While this does enable demining teams to work more efficiently, deploying differing modes of sensation and access also changes how humans understand and relate to the dangers of the ecology. Moreover, these affective differences produce communicative shifts in connection with the explosive ecology. The contingency and weird violence of this ecology, the overarching state of deadly ontological indeterminacy that characterizes human interactions with explosives, as described earlier, lose some of their epistemological uncertainty as trust in the sense structure of dogs communicates a different degree of epistemic confidence and, consequently, a different emotive relationship to the demining practice. The sense of exposure, of vulnerability, of being addressed as a being that can be killed because of its own perceptual limitations, becomes more open. In Spinozan terms, by working with dogs, humans augment their own capacities and, in doing so, introduce the possibility of joyful demining. By allowing dogs to do the work of sniffing, by becoming an animal that follows, human deminers learn paths and practices to interact with the explosive ecology that evade and remove the possibility of experiences cloistered by confrontations with debility and death. Potentiality also accordingly changes from a menacing, uncertain eruption that preys on the human sense apparatus into a more open horizon. As a result, new responses to the work of demining emerge.

Finally, joy alters the work of humanitarianism in larger, more complex ways. Demining operations work as a counterpart or complement to necropolitics. They are a means of reducing the harms of imperial warfare, but primarily by minimizing its lingering violence. In this sense, demining challenges only particular practices of the necropolitics of the explosive ecology. In many respects, deminers and demining dogs are subjected to this same necropolitical power because their lives are exposed to death. Yet, the dogs engage this temporally and spatially distributed war zone not as a site of death making but as a strange series of intensities and encounters with withdrawn things. Joy emerges from this partial access and movement: it reveals a different mode of engaging the violence mired in ecological relations; it is fueled by work, communication, across the incommensurable. Instead of letting violent power define the terms of the space, dogs produce a more porous relationship with the explosive ecology. What is fascinating is that this joy has little to do with orthodox forms of humanitarian compassion. Instead, the dog's joy seems to be a response to sensorially rich encounters with explosives and other objects that make up an ecology of cooperation with human companions. Humanitarianism taps into this process to make demining operations more functionally effective. It is noteworthy that canine joy is not the overt sentimental version typically found in humanitarian literatures. Rather, it involves an openness or interest in the possibilities that presupposes a different relationship with potentiality in the explosive ecology, a product of a distinct constellation of interacting with human and nonhuman otherness. In contrast, humanitarian compassion is often structured by an anxiety about weird violence. This anxiety is anchored in the horrific duality of both knowing that the explosive ecology exists, that it maims and murders, and the epistemological uncertainty of when and how these effects materialize. Sadness develops from the staggering damage of explosives, their seeming inevitability, and the limitations of human agency, despite its vast resources, to stop these dynamics. Sadness is an important response; the point is not to dismiss it, but humanitarian care primarily seeks to reduce this anxiety by removing the explosives, to resolve the ontological contingency of the minefield through recourse to epistemological surety. But, in doing so, it becomes anchored in sadness and horror, and moreover, when underresourced, it fights a perpetually losing battle against contingency. As a result, the potentials of the explosive ecology come to be defined solely by necropolitical power.

In contrast, dogs do not engage these potentialities in the same way. Their mode of care involves an interest, intrigue, and an augmentation of

capacities as they encounter an explosive ecology. Anxiety is not the defining feature of the encounter but the exploration of sensation. While the
net effect of this sensory interaction is the reduction of lethal violence, the
implicit presupposition of this interaction is the ability to linger with the
strangeness and excitement of the potentials emerging from this ecology.
Put differently, dogs clearly do not treat human principles or empathies as
the normative frame or ground for ethical action in this context. Rather,
dogs appear to embrace the joy of strange encounters—following trails,
taking forks, sniffing the ground, and collaborating—in the work of finding
scattered objects. When humanitarianism proclaims that compassion for
other humans is both a good and a means to the good, demining becomes
a sad task, an ever-present reminder of the worst. In this model, the only
possibility for joy develops from managing risk in the hope of alleviating
future danger. Humanitarianism, underneath compassion, transitions
into a practice of securitization that quickly aligns with a technocratic
politics that views the explosive ecology as a functional impediment to
the good life.[100] Framed in this way, grief for the death and injury of the
explosive ecology finds resolution in the numbness of reducing populations to bare life in the process of working to reduce the lethality of the
surrounding environment.

Dogs reveal another possibility: humanitarian practice consists of
weird encounters; playful expeditions; and new meetings between dogs,
humans, and other nonhumans that cannot be fully synthesized by human
sense experience and, moreover, where normative human perceptions,
power relations, and even embodiment constitutes an obstacle to the
emergence of more creative means of contesting violence. In short, humanitarian demining is human, all-too-human, in the way it describes its
practice and understands the structure, form, and value of care. Dogs
exhibit a different degree of generosity in their ecological encounters, one
that produces joy and, in doing so, strengthens the possibility of demining
processes. It is almost as if dogs help to peel back a layer of horror, revealing
underneath its destructiveness a series of uncanny contingencies.[101] They
show that explosives are fully ecological. Although dogs are placed into
this service because of the imperatives of anthropocentrism, at a certain
level, the dogs refuse to be defined by this framework. Their encounter

with others is demarcated not by an imagined fragility and hesitation but by a complex bumping in to the potential otherness of the explosive ecology itself. In doing so, dogs not only subvert their reduction to instruments of anthropocentric reason but reveal another set of practices ongoing in the background and, perhaps, the possibility of another politics beyond the constraints of dominant humanitarian versions of compassion.

In these odd encounters, dogs embrace, contest, and transform key elements of humanitarian demining operations. Their presence as deminers reveals the proximity of military knowledge (and violence) to humanitarian endeavors. In addressing the detritus and slow violence of imperial warfare, dogs form favorable laboring companions. Yet, dogs also illustrate the complexity of these processes. Explosive ecologies are not stable places but horror zones that may drift into the lives of their victims as a condition of a daily walk, as part of passing through an unpleasant environment or even navigating a beautiful one. The violence of explosive ecology relies on masking the longevity of militaristic violence but also on the contingent powers of the environment, minerals, objects, and life-forms. The horror of minefields emerges because of these forces; the necropolitical calculation that some lives may be exposed to explosives; and the twisted, ecological entanglements that enable explosive violence. Dogs help to reveal this background condition of weird ecology as source and site of violence; they demonstrate the martial presuppositions of demining labor and the degree to which land mines are extensions of violence into the colonial present. However, where human demining efforts traditionally view this practice as a sadness, one composed of danger and death, traps and risks, the dogs interact with these weird ecologies as a zone of joyful exchange. This joy sends ripple effects throughout demining operations. The dog's presence in demining operations not only makes it more effective according to the values of anthropocentric reason and labor efficiency but works on the underlying disposition of humanitarian operations. Rather than the tragic labor of human compassion in the face of danger, the scene of humanitarianism becomes one of dogs enacting vivacious engagement and seeking palpable forms of multispecies encounter. Whereas humanitarian discourse is framed by sadness, it draws sustenance from dog demining because of the dog's capacity for joy in response to the strange, weird, dangerous world of the explosive ecology. If humanitarian organizations embrace the dog as a companion in the work of humanitarianism, then they should also consider the dog's interactions as generative of different values, values that pose critical questions about the practice and orientation of humanitarian operations.

2

Heroes, Rats, and the Predicament of Justice

In September 2020, Magawa became the first rat to receive an award for humanitarian work when the People's Dispensary for Sick Animals (PDSA) presented him with its Gold Medal for "devotion to duty."[1] In its announcement, the PDSA praised Magawa for "discover[ing] 39 land-mines and 28 items of unexploded ordnance to date . . . [and] during his career [helping] clear over 141,000 square metres of land (the equivalent of twenty football pitches), making it safe for local people."[2] Magawa's medal came after nearly four years of fieldwork. The announcement was unexpected, and major news outlets, including the BBC, the *New York Times,* and CNN, dedicated articles to Magawa's story.[3] In many of these installments, Magawa's award is explicitly contrasted with the fear and hatred of rats. As one commentary notes, "Not since the fictional Remy of the 2007 Disney-Pixar film 'Ratatouille' has a rat done so much to challenge the public's view of the animals as creatures more commonly seen scuttling through the sewers and the subway."[4] Even when he died in early January 2022, months after his retirement from land mine de-tection, Magawa's life became a sign of the humanitarian potential of rats.[5] These kinds of discourses illustrate that rats require an additional justification to explain their humanitarian potential and to combat the popular sentiment that rats are unseemly creatures. This added justifica-tion demonstrates how implicit presumptions about animality and kinship impact subsequent constructions of humane and inhumane conduct. For instance, for many audiences, it is comparatively easy to make a case that dogs have empathetic capacities and useful abilities, if not underlying humanitarian tendencies. Coevolutionary entanglement combined with

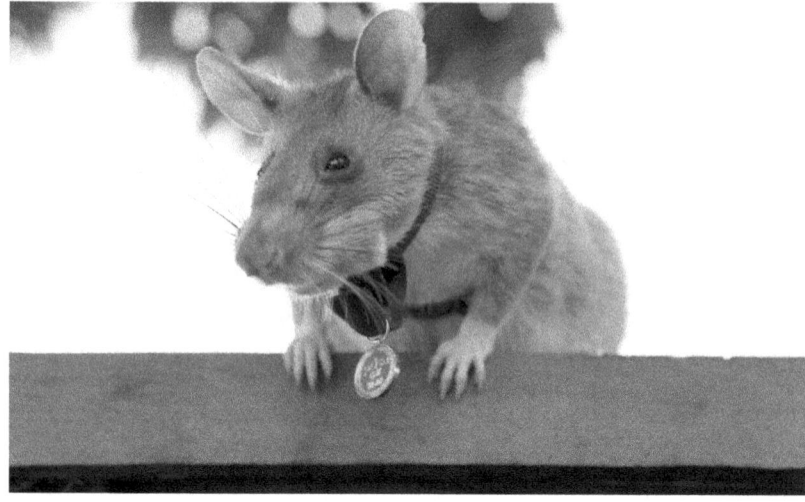

HeroRAT Magawa wears his People's Dispensary for Sick Animals Gold Medal.
Copyright PDSA/APOPO.

cute imagery makes envisioning dog labor as humanitarian labor relatively easy.

Rats are not so fortunate. They are historically despised as the bearers of plague[6] and accused of being a destructive, invasive species.[7] Rats have been constructed to signify the filthy, the dirty, the impure, and the disgusting.[8] Calling another person a "rat" has been understood as an insult for at least five hundred years. The reputation of rats has undergone such sweeping change that Aristotle bemoaned their troubling presence in farm fields nearly twenty-four centuries ago.[9] To this day, military bases and oceangoing vessels bring cats to reduce the populations of rats. Indeed, seventy-one years before Magawa received his medal from the PDSA, the organization formally recognized a cat named Simon with the Dickin Medal for bravery. Simon's primary feat? Preventing the spread of rats, especially one particularly large rat troublingly nicknamed "Moa Tse-tung," aboard the HMS *Amethyst* after it was hit by artillery fire in the Yangtze River.[10] Humans continue to revile rats, to scream and flee at their appearance in their yards, attics, and streets. Pest control products designed explicitly to exterminate rats are a global commodity. Figures of the rat constitute the nemesis of cultural and aesthetic productions ranging from Martin Luther's denunciation of the pope to Tchaikovsky's *The Nutcracker*.[11] As Frederick Wertz puts it, rats are frequently "the target not only of aversion, but of all manners of persecution and extermination."[12] The rat is an animated villain, a plague machine, an infestation, or, at best, a pest, a scavenger, a nuisance, a parasite. If dogs are seen as "humane" because they are nearly human in their capacities and their close kinship with humans, then rats are "inhumane" because they scurry across the line between nonhuman animal and inhuman monster.

This distaste is misplaced. Rats remain part of the fabric of almost every human-influenced environment on earth. The biomes, behaviors, immunity, health, brains, and sociality of rats are close enough to humans' that rats pave the way for observations about human health and behavior.[13] Many of these analyses indicate that there are strong biological and social continuities between rats and humans. Rats live intricate, multidimensional emotional lives.[14] They are increasingly identified as individuals that engage in "complex social behaviour, . . . [and that] have come to be seen as more 'human,' even in the laboratory."[15] In addition to their social and emotional lives, rats have demonstrated that they possess metacognitive abilities and can reflect on the adequacy and inadequacy of their own knowledge.[16] Rats, it seems, employ reason alongside their emotional and social skills. Furthermore, although they are not as "cozily" coevolved with humans as dogs, rats and humans share an intricate, interwoven history.[17] As several distinct species, rats are massively distributed geographically, culturally, and historically. Ancient fields have rats. Urban superstructures ooze with rats. Rats sail across the seas. As you read this, there are rats flying on planes and possibly on a space shuttle.[18] There is almost no arena of human habitation that has not been formed through an interaction, often unconsciously or unintentionally, with rats. In this sense, rats exist in both mutualistic and parasitic relation to humans. As Michel Serres shows, the very distinction between parasitism, mutualism, and sociality breaks down.[19] Consequently, rats, as with dogs and other nonhuman animals, exist in multiple relations with humans and with one another.

Despite the complexity of these relationships, dominant attitudes toward rats view them as the antithesis of humanitarian: they are base rather than cultivated, they are selfish rather than altruistic, they bring miasma rather than health, and so on. Many of these perspectives hinge on the potential for rats to disrupt practices of agriculture and consumption by spreading disease and spoiling stores of foodstuffs. In this sense, rats are an annoyance because of assumptions about the proper ordering of human life designed to produce excesses, surpluses, or reserves. Of course, the existence of surpluses inevitably attracts nonhumans that can also live off these stores. The disgust with rats emerges as a response to a sentiment about the moral economy of agricultural development. The distaste also exists in sharp contrast with the appearance of rats in humanitarian affairs and their capacity for humanitarian labor. Magawa's work is a component of an assemblage of rats, humans, and technologies that has become a nascent aspect of humanitarianism. Specifically, Bart Weetjens, founder of APOPO, first realized that rats, with their strong olfactory capacities, could plausibly assist in the detection of mines and explosives, much like dogs. Since his initial intuition in the mid-1990s, Weetjens established APOPO and began to test, breed, and train African giant pouched rats *(Cricetomys ansorgei)* to form an efficient, inexpensive land mine detection force.[20] Magawa is the most successful of several generations of detection rats bred, qualified, and cared for by APOPO. Since they began to work in explosive detection, rats have greatly expanded their roles in humanitarian work to include infectious disease diagnosis, detection of illicit wildlife transfers, and care and curiosity in their interaction with humans.

The introduction of rats into these roles involves different challenges than for dogs because rats exist in a liminal zoological and political space: effective at humanitarian work but confronting much more consolidated, if culturally specific, phobias. As such, exploring the role of rats in humanitarianism offers a different account of the exclusivity of humanitarianism because rats, while empirically useful at specific tasks as judged by anthropocentric reason, encounter alarmingly and explicitly inhumane responses in other contexts. Humanitarian work with rats thus involves multiple intersecting challenges: it seeks to capture and employ the rats' capacities to expand opportunities for human habitation and flourishing; it endeavors to increase the set of technologies available for humanitarian service; and it contests a particular mode of anthropocentrism, a fear of rats' parasitic potential, to qualify them as humanitarian.

In addressing humanitarian rats, this chapter explores the conditions of possibility for humanitarianism as a mode of practice. If the problem facing explosive detection dogs is that the recognition of their abilities as humanitarians is predicated on a kind of misrecognition that leashes them to the whims and sentiments of anthropocentric reason and anthropocentric feeling, then rats, in contrast, confront a different problem, namely, being constituted as an enemy of humanity, even an enemy of all. Indeed, if, as Daniel Heller-Roazen suggests, the first formulations of the category of humanity rest "entirely on the possibility of isolating, in human beings, an abstract principle of species, humanity, which is to be 'used' in accordance with its intrinsic 'worth,'" then the rat constitutes the antithesis of this principle of species as a threat to humans understood at the level of their biology, of their species-being, or their mode of collective existence.[21] However, underneath this sense of collective existence is an attachment to a particular model of organizing human–ecological relationships within agricultural societies as they have been predominantly framed by colonial and Eurocentric norms. As more complex agricultural and industrial systems develop and are exported around the globe, rats transform into an embedded feature of human life across the planet. Disgust and distaste for rats intensify as these conditions spread. Humanitarianism, the chapter argues, is also conditioned by these ecological practices, and moreover, it is the continuity of a paradigm of agricultural power that makes rats thinkable as humanitarian laborers. In this sense, humanitarian interactions with rats, even highly empathic ones like APOPO's approach, reveal that Didier Fassin is correct when he argues that humanitarianism is underwritten by a politics of life, one where universalism and sympathy are predicated on the form of life under consideration and where specific paradigms of sociality and ecology underlie dominant interpretations of humanitarian need.[22] Rats appear as humanitarian actors only after becoming a new subject of knowledge, one that is malleable according to anthropocentric reason and that promises to extend this human life. Their rebirth as humanitarian laborer blossoms from their capacity to function as a technology, a prosthesis of human sensation, but one that is also capable of affection as humanitarian work transforms rats from parasites that expose the fragility of agricultural life by creeping within its crevices to gnaw and chew into the means of clearing land mines, preparing the way for agriculture, and preserving human health.

Rats expose two different limits within humanitarianism. The first limit involves the biopolitical dimensions of humanitarian practice as both the source of fear of rats and the basis for their admission into humanitarian worlds. The second limit concerns the question of whether humanitarianism can articulate forms of care and justice in more expansive terms, stretching beyond what William Connolly calls the "human estate," toward a stranger, more multidimensional interpretation of giving and care.[23] Even though APOPO's work clearly reproduces forms of anthropocentric reason, using animals to reduce a risk to human communities, it also contests forms of explicit anthropocentric violence. In effect, rat humanitarians reveal the ambivalent possibilities of nonhuman humanitarianism. What APOPO encounters in making the rat into a humanitarian is the problem of how to articulate the contribution of rats within an anthropocentric symbolic economy. Animal welfare and formal recognition constitute one way of recognizing the burdens of the rats' incommensurable gifts of labor and sensation. As such, APOPO seeks to afford the rat a unique place in humanitarian work; reveals that rats contribute to humanitarian generosity, a core tenet of humanitarian ethos; and inadvertently shows that these gifts may originate from nonhuman otherness.

This chapter addresses these themes in four sections. First, the chapter explores the emerging research on rats as valuable agents of humanitarianism. Examining the studies of APOPO and other humanitarian communities, it describes the biocultural capacities of rats that make them effective at land mine detection, disease identification, and other humanitarian functions. Second, the chapter illustrates how disgust with rats hinges on specific types of agricultural thought and practice; the chapter demonstrates that rat humanitarians need to overcome not just affective disgust but deeply sowed metaphysical prejudice. In doing so, the chapter also demonstrates how humanitarianism tacitly relies on a highly attenuated vision of human relationships with ecological space. Third, the chapter returns to the rats to discuss how their humanitarian capacities hinge on forms of care and ecological sensitivity that contrast with human perceptions. As such, this section makes the case that rats are singular and strange in their forms of humanitarian outreach but also differ decidedly from the research emphasis of humanitarian organizations. The final section explores the paradoxical status of justice in the operation of rat humanitarianism. It argues that humanitarian pursuits aspire to construct a new model of legal order, but this model reaches an impasse when it encounters rat contributions to humanitarianism. This impasse, however, is not a dead end but a critical moment to reconsider the constitution, form, and trajectory of humanitarian practice.

BECOMING RAT, BECOMING HUMANITARIAN

Bart Weetjens, founder of APOPO, first considered the possibility of rats working as deminers after reading about the scent detection abilities of gerbils in the mid-1990s. According to his own account, Weetjens began discussing rat mine detection with his mentors from the University of Antwerp. These conversations led to the first research grants to explore the viability of working with African giant pouched rats in late 1997.[24] Over the next seven years, Weetjens and the newly formed APOPO experimented with training procedures, tested the efficacy of rats in demining procedures, established breeding practices and facilities for rats, and set up permanent operations in Tanzania. In the early 2000s, APOPO's methods of mine detection training were externally verified by the Geneva International Centre for Humanitarian Demining. By 2004, the first group of mine detection rats were licensed by the International Mine Action Standards, and rat land mine detection became an accredited practice of land mine detection.[25]

Since then, APOPO has expanded the humanitarian functions of the newly branded HeroRATs in several ways. Historically, teams of rats have worked in explosive detection in Mozambique, Cambodia, and Angola.[26] The scientific study of rat scenting abilities improved the precision and efficiency of these land mine detection procedures. In this process, rats became a new subject of knowledge, and APOPO developed several new technologies to assess the health, growth, and functionality of rats as well as their social and emotional behavior. Rat scenting abilities have also been applied to different problems, including detecting tuberculosis (TB); discovering illicit tobacco products; performing search and rescue for persons trapped under debris; discovering illegal wildlife transfers (such as pangolins); and hypothetically identifying other medical problems, such as salmonella and brain disorders.[27] With the exception of TB detection, many of these areas of study are in their infancy but build from the growing evidence of the rats' exceptional olfactory capacities and their consistent rate of successful training. Much like dog detection, rats perform these tasks with the assistance of an assemblage of human support teams and technologies. This assemblage actualizes the potential for rats to find land mines, TB, or other environmental features and introduces the possibility of safer, more efficient humanitarian practices.[28] Finally, the rats have become central to the promotional and organizational success of APOPO. Through a series of advertising efforts, public acknowledgments, and media campaigns, such as Magawa's award, the organization uses "HeroRATs" to brand its signature approach to humanitarianism. It is now possible to watch multiple videos of rats at work demining or to remotely "adopt a rat" to contribute to APOPO's work.[29] In the span of twenty-five years, the possibility of rat humanitarianism transformed from a fascinating thought experiment into a mobile assemblage of rats, humans, organizational networks, medical facilities, and scientific knowledge.

In many ways, the inspiration behind both Weetjens's and APOPO's work reflects what Liisa Malkki calls "the need to help," or a mode of articulation that identifies taking responsibility for the world at large as integral to the construction of the self.[30] Weetjens and APOPO reproduce a set of common tropes and representations about suffering in the context of international humanitarian work.[31] At the same time, the organization frames its work as one of empowerment, of minimizing the long-term reliance of developing communities on intervention from former colonial powers. In one public address, Weetjens describes the vision of APOPO: "You may think this is about rats, this project, but it in the end it is about

HeroRAT Ronin in a minefield in Preah Vihear. Copyright APOPO.

people. It is about empowering vulnerable communities to tackle difficult, expensive, and dangerous humanitarian detection tasks and doing that with a local resource plenty available. So, something completely different is to keep on challenging your perception about the resources surrounding you, whether they are environmental, technological, animal, or human, and to respectfully harmonize with them to foster a sustainable world."[32] Although Weetjens has stepped back from an active role in managing APOPO, the articulation of these principles helps illuminate what first led to the possibility of involving rats in humanitarian practices.

Weetjens's statement is interesting not because it emphasizes empowerment, another common theme in humanitarian genres, but because of how it stresses the importance of exploring the limits of perception in relation to human well-being.[33] There are several ways to interpret this statement. On one hand, APOPO's project develops from the capacity to understand rats as meaningful participants in detecting explosives, to embrace something akin to a flat ontology where the interactions between ecology, technology, and nonhuman and human life generate novel possibilities for political action.[34] In doing so, Weetjens implicitly rebuts traditional humanitarian meditations, which often identify human aspirations, feelings, and rationality as the basis of humanitarian action, and instead locates this potential in shifts at the level of perception and in the engagement with other objects and life-forms. In this sense, Weetjens openly pushes humanitarianism beyond its traditional anthropocentric frames of reference. HeroRATs emerge from this revaluation of perception and values. On the other hand, anthropocentric reason leaves a powerful imprint in this statement because it ultimately expands perception for the purpose of converting ecological difference into a resource as part of a project about securing human welfare. By point of comparison, if Weetjens is compelled to argue that rats can become humanitarians, based solely on this statement, it is far less clear whether he maintains that humanitarianism, in turn, should be organized to promote the flourishing of rats in general. At stake in this emphasis on perception and organicism is a kind of providential view of the world as a set of deep ecological resources for human flourishing. Stretching perception is a far more problematic exercise if it ultimately works exclusively as an extension of anthropocentric reason.[35] Weetjens's position, and arguably the HeroRATs project, thus exists in an ambivalent space between advancing the perceptual, ethical, and symbiotic dimensions of humanitarianism and reaffirming human primacy.

Rats have several capacities that preadapt them to this form of humanitarian labor. African giant pouched rats are long, thick-bodied rodents and the main nonhuman animal trained for the detection of land mines.[36] Larger than many traditional rat subspecies, scientific debate exists about whether *C. ansorgei* is a "true rat" or a different, larger cousin rodent.[37] While other rat species, including the Columbia Wistar rat and even white rats, are being experimentally trained for land mine detection, African giant pouched rats boast a profile of twenty years successfully demining and identifying infectious disease. This success depends on two key abilities. First, these rats possess uniquely strong olfactory senses.[38] Like dogs, rats are equipped with cilia hairs that facilitate the sensation and differentiation of odors from passing gaseous particles.[39] Rats also evolved specialized anatomy, called glomeruli, that activates in response to distinct smells. Glomeruli play a role in stimulating variable patterns of neurological activity in rats when they are exposed to distinct odors. This burst of neurological activity indicates that rats may have additional aptitude or affinity for the act of distinguishing the components of a smell irrespective of what the act of smelling encounters.[40] Put differently, the rats may simply enjoy exercising the faculty of smell in a way that is different from humans and other nonhumans. Like dogs, rats have a vomeronasal cavity that traps odor particles in mucus to improve the rats' ability to differentiate between chemical signatures.[41]

Second, African giant pouched rats have notable aptitudes for social learning and training. Like many nonhumans, rats engage in highly interactive social relations. These are nurtured during an extensive period following birth during which growth and development coincide with regular maternal affection and care. Following this period of postnatal development, African giant pouched rats will also readily socialize with humans if the interactions are introduced during an early enough phase of their development.[42] APOPO's methods for training mine detection rats involve a deliberate breeding process and careful monitoring of the process of interspecies interaction. Breeding rats with an aptitude for tameness, strong scenting abilities, and an aptitude for trainability, new generations are then given an appropriate period of gestation with their mother before human–rat socialization starts. Over time, the rats accept human companionship and appear to find comfort in the relations they form with their handlers. Once these relationships are established, the rats begin a carefully sequenced training program. This process begins by teaching rats to differentiate and consistently detect chemical traces of TNT (explosives), before proceeding to field testing with unburied, then buried devices, before a full "live" trial in APOPO's own facilities with various types of real defused land mines.[43] During each stage of the training, the rats learn through positive reinforcement techniques, and they are held to exacting standards to ensure the quality of their work. In the case of land mine clearance operations, the rats are also too light to trigger most explosive mechanisms. They are able to move easily through areas too small for normative modes of human ambulation and are comfortable working in multiple ecologies. APOPO's research, which has faced some challenges on scientific or behavioral grounds, maintains that "rats can search around 200 square meters in 20 minutes. This would take 25 operational hours using metal detectors (up to 4 days). [Rats] can detect both metal and plastic-cased landmines, making them highly efficient landmine detectors."[44] From the perspective of anthropocentric reason, this rate of detection, combined with their relatively inexpensive training, diet, and medical care, makes rats a far more efficient and precise mechanism of mine detection.

The rats' olfactory and socialization capacities have also been experimentally repurposed to address other problems. These include infectious disease identification, in particular the identification of TB; detecting the presence of tobacco to assist in the illegal sale of cigarettes and other tobacco products;[45] or finding salmonella in the fecal matter of horses.[46] Studies have further demonstrated that rats may have the ability to search for humans impacted by a collapse of the built environment.[47] In this case, following an event like an earthquake or explosion, rats equipped with video cameras may be able to enter otherwise inaccessible debris to search

HeroRAT Ellie gets a reward. Copyright APOPO.

for survivors, vastly accelerating the time frame of search-and-rescue operations. In each situation, the success of rat humanitarianism hinges on the rats' capacities for scent detection, size, and comfort with humans to identify distinct signatures in each environment, but these abilities, adaptability to multiple contexts, efficiency, and lack of expense emerge as by-products of evolved capacities and cooperative work with humans and other technologies, which transform the rats into a valuable part of humanitarian work.

This description may give the impression that APOPO views rats solely as a tool. However, the organization pairs its management and development of HeroRATs with an explicit articulation of an ethos of "animal welfare," veering into the terrain where humanitarian and anthropocentric sympathies resonate with one another.[48] This ethos leads to several practices intended to provide the rats comfort and happiness, including building kennels and play cages for rats that enable them to socialize; offering a variety of play activities, such wheels, toys, and puzzles; providing spaces for exercise; and supplying the rats a regular diet of fresh fruit, fresh vegetables, nuts, sardines, and rodent pellets. HeroRATs also receive routine checks by veterinary services that specialize in the health of rats, and the organization's main facilities include a separate sick bay facility to assist rats impacted by infection, parasites, or other health conditions. At around six or seven years of age, the rats retire and live out the rest of their lives in a similar facility, where they continue to receive their typical diet, play opportunities, and health care until end of life.[49] Both the biological well-being of the rats and the quality of the rats' lives are key points of emphasis in APOPO's articulation of animal welfare.

APOPO's emphasis on animal welfare serves other discursive functions as well. As with detection dogs, the proximity of the rats to the danger of explosive precipitates as well as domestic and international animal welfare laws (depending on context) create a need to produce assurances about the rats' welfare. As the introduction argues, humanitarian sympathy and anthropocentric feeling reinforce one another because both ascribe a kind of innocence and powerlessness to the others they serve.[50] At this point, the notions of animal welfare and animal rights are perhaps the most dominant frameworks for the protection of nonhuman animals because it emphasizes the similarities between humans and animals with respect to pain, reason, and making the latter de facto subjects of rights and law.[51] These concepts use an anthropocentric paradigm of autonomy and sovereignty to articulate the needs of nonhuman animals (animal rights movements similarly employ anthropocentric terms), which minimizes the political significance of the differences between nonhumans and humans. In short, it functions, much like humanitarianism, by making nonhuman animals appear more human. However, as Zakiyyah Iman Jackson persuasively argues, the practice of humanization may not benefit so-called nonhumans or what she terms "blackened bodies," liminal to the category of the human, because the very plasticity of animality and humanity, their points of contact, ultimately legitimates both care and torture.[52] APOPO, as an organization, remains indebted to humanitarianism's broader conceptual lineage and articulates its version of rat welfare in a way that reproduces many of these features of animal welfare discourse. In a sense, animal welfare misunderstands the depths of rat agency in making explosive detection a problem of shared work across species difference. Indeed, it would be a mistake to read rat capacities solely as a product of innate traits of rats that build from instinctive patterns of behavior. As Vinciane Despret demonstrates in the context of laboratory work with rats, an entire mode of world building, a model of ecological awareness, and nonhuman thought subsist in rats' interactions with humans and their surroundings.[53] Rat agency is not a result of default capacities but molded in relation to the problems they engage and the others they encounter. Land mine detection or infectious disease diagnosis with rats involves an entanglement of individual rats, the species of rats humans interact with, and the rats' *Umwelt* as it develops in relation to others. Shifting the framework to appreciate the rats' capacities as generative, agentic, and perspectival helps to explain why rats can become humanitarian laborers but also requires a politics beyond basic questions of animal welfare because it raises a question about the depth and quality of ethical relations beyond care for mere existence. Though these insights can often be extended to nonhuman companion species, rats face much more intense opposition because they are typically subjects of anthropocentric disgust. Understanding the success or limitations of APOPO's involvement with rats, and the problems the organization encounters by introducing rats into humanitarian domains, hinges on understanding why disgust and fear of rat agency are so culturally potent. Put differently, hatred of rats is old—older than modern epistemological distinctions articulating the distinction between humans and nonhumans in terms of species life. Rats are rarely given the opportunity to be seen as benefactors of human societies, let alone complex forms of worlding. Unpacking these conditions of possibility helps to explain why elements of APOPO's work with rats constitute such an important development in multispecies justice, even as its work rearticulates dominant paradigms of animal welfare.

BASTARD RATS

APOPO's use of animal welfare is not the first time this discourse has been applied to rats. Brown or "fancy" rats were first domesticated as pets during the late nineteenth century. This process of domestication created a small constituency of humans who advocated for rat welfare.[54] However, the dominant perception that rats were pests, spreaders of disease, parasites, and, at best, nuisances continued to grow. These reactions contain a grain of truth in terms of the jeopardy rat–human interaction can pose for human well-being. For instance, black rats were arguably responsible for the historical movement of multiple infectious agents into different geographic communities. As historian William H. McNeill maintains, the homogenization of urban housing in Europe and the vast increase in shipping during the rise of capitalism made it possible for rats "to enter a new ecological niche that permitted them to spread beyond their original homeland."[55] Rats were recognized as a vector for disease prior to the identification of viruses, bacteria, and other microbes as sources of pandemic. Nonetheless, even after the invention of contemporary biological and medical science, and serious disputes over the veracity of this explanation, rats still engender responses of disgust and loathing.[56] The depth of this disgust is perhaps most evident in the fact that many episodes of genocide and gratuitous violence involve labeling specific communities as rats, a comparison that legitimates their extermination. One way of interpreting this response to rats, as well as the persistence of such overt phobia over multiple centuries, is to situate the animus in a broader set of social conditions. Examined as part of a larger milieu, rats appear as culprits because of a group of political, ecological, and ontological anxieties about health, precarity, and finitude alongside their association with epidemic.

Disgust with rats typically occurs in the context of eating and agriculture. As Sarah Ahmed argues, disgust is an affect that emerges as a response to the ingestion, or thought of ingestion, of something untoward.[57] Rats regularly gnaw on foodstuffs, spoil supplies, leave droppings, and create a form of ecological contact that is perceived as viscerally unsettling because of the potential to spoil objects of consumption. Disgust arises from this possibility of contamination, which problematizes the boundaries between the clean and the dirty, the healthy and the ill. These boundaries, in turn, develop from a set of commitments produced by specific modes of agriculture, consumption, and ecological relations. In their work on dark ecology, Tim Morton offers a useful heuristic for interpreting this form of political ecology, which they term "agrologistics." Agrologistics defines a group of metaphysical commitments, aesthetic imperatives, and moral principles that structure agricultural societies.[58] These commitments form a consistent pattern or model for governing the relations between humans and other objects.[59] In particular, they produce anthropocentric prejudices with respect to ecology difference. More specifically, Morton contends that agrologistics treat the strange, partial, and twisted character of ecological relations as a field, an environment designed to be governed for a specific end, namely, the perpetuation of human existence. Agrologistics grows from three implicit presuppositions: first, that the law of noncontradiction is inviolable, thus each animal, vegetable, and mineral has a clear and distinct identity separate from others'; second, that existing means constantly being present, or, put differently, existence has no meaning if it is withdrawn or absent; and third, that existing is always better than any quality of existing, so the perpetuation of life becomes the ultimate purpose of a life.[60] These principles explain several recurrent features of agricultural societies, such as the persistence of hierarchal power relations, the segregation of human (and nonhuman) labor capacities, the fetishization of patriarchal domination, the invention of a nature–culture distinction, and the conversion of nature into a set of resources for the development of culture.[61] Here anything that enables human existence to fend off contingency, including the exclusion and elimination of forms of human life, becomes a resource, while all things that harbor the capacity to trouble stable distinctions transform into weeds, dangers, and pests that jeopardize agrologistic fantasies. For Morton, agrologistics works as kind of slow-motion catastrophe that grows into other disasters, such as global warming, the Industrial Revolution, and racist massacre, each of which takes agrologistics as its condition of possibility.[62]

From the perspective of agrologistics, the rat is a parasite, an in-between creature, both threatening individual and crawling horde, which can enter and eat in the fields but has no place there. Rats eat crops; they interfere with designs and bring infection. These dangers (pestilence, miasma, spoilage) only appear as dangers because they problematize the distinction (spatially and ontologically) between the well-tended field and weird ecology. They are problems of boundary, of the threatening character of things that lie in-between or that never fully reveal themselves, playing with the distinction between presence and absence.[63] Centuries after the development of agrologistics, rats continue to incite fears just as they produce concrete dilemmas for human societies by consuming grain or thriving in basements and insulation. In a sense, rats jeopardize all three of the principles of agrologistics: they undermine the persistence of human life by eating and destroying surplus (the reserve promised by agrologistics that guarantees the persistence of the present into the future) and by spreading infection (bringing death); they refuse to reveal themselves or, put differently, to be present (although humans know that they eat and crawl, they do not offer themselves as adversaries and slip into the fields using rhizomatic paths and tunnels); and they blur the distinction between the field and the outside, snubbing what Deleuze and Guattari might call agrologistics' "plane of organization" in favor of a "plane of immanence."[64] It is rats' ecological otherness, their entanglement in a partial set of disruptive, now planetary-scale interrelations, nesting within agricultural habits, that constitutes their being as a threat to the agrologistic project and, consequently, that provides an explanation for the recurrent disgust directed toward rats across seemingly disparate temporal and political contexts.[65] Rats remain a virtual scourge for agricultural societies because they create problems for everyday living, but these problems emerge because of the instability of the metaphysical assumptions planted deep within agrologistics.

The hatred of rats persists as a remnant of this reaction because agrologistics remains a formative force within contemporary capitalism. As such, rats threaten the productive, domestic, urban, and good life.[66] Despite their limited domestication, they constitute an other to agriculture, industry, and housing as they devour crops, eat into plastic, chew electrical wires, burrow through trash, or take free rides across the oceans. Rats crawl through attics, sewers, and cellars. They scramble across city streets and interfere with the shipping and storage of goods. The danger rats pose to the field (and therefore agrologistics) has thus not disappeared so much as multiplied and differentiated. Rats continue to be a threat because they are parasitical on capitalist flows of interconnection.[67] As a consequence, an entire industry of pest control products and rat extermination services seeks to manage and eliminate rats from human societies. For instance, rat poison is discursively identified as targeting rats in their species-being, even though the chemical causes internal hemorrhaging in any warm-blooded mammal (depending on mass and quantity of chemical) and unintended harm and ecological damage.[68]

If disgust with rats emerges from the metaphysical commitments of agricultural societies, then the construction of ever more complex assemblages of political, industrial, and economic life based on these commitments will be unable to address disgust. Rather, the expansion of agrologistics will amplify the spectral quality of rats until they become subject, as Neel Ahuja might put it, to a "government by species."[69] Although affects of disgust with rats are, to a degree, culturally specific, the emergence of dominant practices of agriculture, boosted by colonialism and capitalism, catalyzes disgust with rats across different zones of time and space. These modes of disgust, in turn, colonize other possibilities for ecological politics and become such a powerful discourse that disgust with rats becomes the symbolic means of excluding and exterminating human communities.[70] Indeed, as Rafi Youatt puts it, "a multi-century history of rat extermination efforts following both plague and pestilence . . . testif[ies] to the direct impacts that rats have had on global political life. . . . The concept of rats as reviled, diseased vermin to be exterminated has been politically transferred onto a variety of human populations who have been tagged as rat-like, as humanrats, or simply, as rats, with deeply negative consequences."[71] In contrast, the rise in positive sentiment toward rats and domestication of specific rat subspecies historically occurred once new technologies enabled control over aspects of rat behavior such that they no longer posed a threat to agrologistics. In other words, rats become companionable precisely because they are "captured," in Antoine Traisnel's sense of the term, by contemporary epistemologies, and their life transforms into a problem of governance, deprived of their ecological weirdness, subject to the prejudices of identity, the being-present of life in a cage, terrarium, or laboratory.[72] Yet, rats continue to frustrate agrologistics as an excess, even as some rats become the means to extend agrologistic projects. Youatt aptly describes it thus: "Rats point the way to the limits of human life as a sovereign, territorially bound life—geopolitics sees ecology primarily as a resource question and the conditions necessary to sustain extraction, but rats show some of the limits of that by the inherent commensalism that other species will undertake. . . . Rats highlight the sometimes-unwanted dependencies that occur as an inherent component of that condition."[73] Agrologistics produces several different positions for rats as subjects of laboratory control, disgust, invasion, and extermination.[74]

THE PARASITES OF ALL

Reading the history of rat phobias through the prism of agrologistics helps illustrate a duality within humanitarianism. On one hand, humanity is demarcated by distancing itself morally and ethically from its constitutive outside: the inhuman. The figure of the inhuman, such as the rat, is characterized by selfishness, baseness, and injustice, the cluster of traits that oppose human civility, decency, and respect, but also the blurring of boundaries, the failure to appear at hand, and the destruction of reserves. On the other hand, the HeroRAT functions as a humanitarian laborer because of a shift in the perception that principally concerns ecology and aesthetics. As the first section of this chapter described, HeroRATs work as part of an assemblage; they are bred, trained, and assisted by human attendants, but their human caretakers have also molded and trained themselves to attend to their rats, so that they mutually enhance one another's protection. This process requires humans to work on themselves as well, to become capable of interacting with the rat's world, to encounter an ecological sensibility that is not their own. It also demands that rats undergo a social metamorphosis to become capable of being handled, touched, and led. While dogs are often so thoroughly immersed in human worlds that their "wildness" gets subsumed by cohabitation, rats are not so easily integrated into humanitarian frameworks because they remain parasitic to agrologistic endeavors.[75]

The passage from rats as parasites to rats as humanitarians occurs, as the previous section argues, because of a shift in perception. The question concerning HeroRATs is whether this shift is one internal to agrologistics, making the frightening figure of the rat into something unthreatening, cute, and ultimately beneficial in the extension of agricultural politics, or, conversely, represents a break with the anthropocentric views of nonhuman life traditionally at work in humanitarianism. In one sense, subject to scientific knowledge, controlled by a variety of technologies ranging from behavioral analysis to brodifacoum, rats become instruments of agrologistics as they reduce the danger of minefields and clear space for agricultural production. Put simply, the domestication of rats, either as pets or as subjects of science, is an agrologistic solution to a problem engendered by agrologistics. However, in another sense, it is from within these conditions that new encounters between rats and humans also take place. Rather than understanding the "capture" of rats by scientific insight or experimentation exclusively as technocratic control over species life, it is valuable to think of this as a mode or zone of interaction, even if this interaction is clearly marked by hierarchical forms of power.[76] As Bruno Latour contends, the laboratory is often the site where scientific encounters not only influence politics but constitute political activity itself.[77] So, too, the enactment of scientific engagement and close encounters with rats shifts the mode of access between humans and their rat messmates, at times introducing new sensuous and ethical possibilities.[78]

APOPO's engagement with rats thus both reproduces and breaks with the duality at work in other humanitarian versions of agrologistics. Rats become thinkable as humanitarian laborers because they behave according to the desires and dictates of anthropocentric reason. They decrease the danger of minefields and disease in ways that are knowable, controllable, and for human benefit. Here epistemology, in combination with cohabitation, transitions the rat from a danger to the well-tended field to a cure for the plague of war. Deprived of their threatening character, rats reproduce agrologistics in the case of demining by clearing land to enable farming and a return to a normative image of social life. Indeed, one of the common arguments regarding the long-term impacts of minefields is that they undermine the possibility of agriculture and industrial development.[79] Agrologistics is, in many respects, opposed to the deterritorialization produced by warfare and the ontological indeterminacy of the explosive ecology explored in the last chapter, where contingency and violence undermine the creation of striated, settled fields and hierarchies.[80] Successful land mine clearance paves the way not only for greater safety but for the agricultural life. Rats are converted into humanitarians because they no longer pose a threat to a field overdetermined by the dangers of land mines, as scientific understanding transitions their parasitism into an object of governance and as anthropocentric feeling creates more optimistic impressions of mammalian others through interaction. As a contributor to agrologistics, the rat suddenly becomes intelligible as a humanitarian laborer. Anthropocentric feeling further tugs HeroRATs into a genre of innocence so they become affectively imbued with the virtues of a humanitarian.

Yet, APOPO also articulates a vision of rats as meaningful partners and not simply anthropocentric props.[81] In its materials, APOPO repeatedly centers the rat as a key agent in the humanitarian process. Rats "are helping to find landmines and detecting tuberculosis as part of an integrated approach. . . . [The rats] speed up the process, getting people's lives back on track as fast as possible."[82] Here the rat is framed not as a mere tool but rather as a collaborative partner in detecting and removing explosives. It is important to take APOPO's rhetoric here seriously and not simply as an ideological maneuver or marketing strategy. Though APOPO's partnership involves the use of rats and is designed to improve the welfare of human communities, these commitments performatively identify the quality and consistency of the rat's life as an important consideration.[83] Care for rats becomes an implicit part of APOPO's organizational mandate, just as explicit attention is given to the health, quality of life, and longevity of HeroRATs. Multiple features of APOPO's involvement with rats differ markedly from other, more overtly anthropocentric models of humanitarianism. As the next chapter shows in detail, if HeroRATs expand agrologistics, they do so with considerably less violence than other forms of nonhuman humanitarianism.

Prior to the incorporation of APOPO, Weetjens and the other members of the organization entertained the possibility of rat detection operations because of rats' notable scenting abilities. While identifying the distinctive olfactory capacities of rats opened the possibility of new practices of

Magawa, wearing his PDSA Gold Medal, with his handler, Malen. Copyright PDSA/APOPO.

explosive detection, cohabitation with rats was also an important precondition for this insight because APOPO's approach to training HeroRATs relies on a series of social encounters and the development of a mutualistic relationship facilitated by the cohabitation of humans and rats. In making this shift, APOPO confronts the ecological complexity not only in raising rats but in the interaction of its members (humans and rats) with explosive ecologies, environments where the ontology of armed conflict clutters the supposedly clean lines between culture and nature. Rats could be "adapted" to the minefield because of the proximity and sympathies they recently established with humans but also because the explosive ecology is a terrain or object where orthodox, agrologistic models of human life cannot persist. The explosive ecology opens possibilities for becoming and interaction that the implicit injunctions of agrologistics would otherwise prevent. For the rats to be thinkable as a resource, specific humans also need to socialize with and come to appreciate the rats' engagement as a different form of the partial interaction with other ecological objects. Learning not only to think but to imagine the sensation, feeling, and interaction of the rat was, APOPO maintains, a crucial transformation.[84]

Species difference also constitutes a means of demarcating different approaches to land mine clearance work itself. As Darcie DeAngelo brilliantly argues, APOPO's inclusion of rats in detection labor breaks with martial themes common to some areas of land mine detection.[85] As DeAngelo notes in the context of APOPO's work in Cambodia, rats' involvement precipitated new discourses that viewed them simultaneously as objects of endearment and affection, effective tools, and disease-bearing invasive species.[86] The rats' cuteness, their involvement with multiple human land mine detection teams, and their existence outside of the longer discursive history linking dogs, military practice, and demining work created breaks that tugged against the nationalist and xenophobic tendencies of some forms of mine detection. Rats thus disrupt, through their involvement, affective impacts, and symbolic significance, many of the resonant features of militarism and humanitarianism. In this sense, labor with rats also constitutes an intervention internal to humanitarian organizations on what DeAngelo calls the "ethos within an organization given its historical and political contingencies."[87]

The contentious role rats play in shifting humanitarian practice helps illuminate the limits of traditional accounts of humanitarianism that view intrinsic human capacities, human suffering, and philanthropic gestures as universal and apolitical. In this model, it is humanity that is the subject, origin, and judge of the humaneness of an action and determines the inhumanity of the enemies or traitors of these ideals. The explicit and implicit contestations DeAngelo documents illustrate how nonhuman animals also contribute to struggles within the terrain of humanitarianism over the partnership, practices, principles, and perceptions that matter. Work with rats, as a novel nonhuman partnership in humanitarianism, builds, as Weetjens's interview states, from a "*shift in perception*" to understand the environment, the objects, and the subjects of humanitarianism differently.[88] These shifts in perception are not intrinsic to human life but assembled from structures of sensation and imagination that develop across ecological and creaturely difference. Put differently, rat humanitarianism could not emerge in the absence of behavioral science, in which all action becomes a means of communication, of new prosthetics of sensation, of the rise of ecological sciences, and so on.[89] Rat–human couplings materialize from shifts in ecological disposition that are not germane to human thought or perception but modulated through the interaction of incommensurate worlds, composed of different sensations and, consequently, different values. Arguably, humanitarian generosity expands not by assuming humans have a capacity for compassion but, as Jacques Rancière might put it, through a new "partition of the sensible."[90] While humanitarianism is often fetish-like in its obsession with new technologies, Katja Lindskov Jacobsen convincingly contends that humanitarian action and principles change substantively as they interact with new prosthetics. As she puts it, "not only does the humanitarian use [of technology] affect the social constitution of a new technology . . . [but] technologies can also affect the construction of social relations—in other words, technology is not only constituted . . . but technology also has a degree of vibrant, constitutive agency."[91] In a broader sense, humanitarianism itself constitutes a type of technology, a set of concepts, practices, and aesthetics of self and social articulation that serve as models to reorient perception around particular outcomes.[92] Even as rats are framed as technology, in subtle ways, they shift the articulation of humanitarian projects. As such, they constitute a plausible case where the interaction of incommensurate worlds, communication about creaturely difference, creates a partition in sensibility.

With APOPO, there are two possible but not mutually exclusive interpretations of this shift in perception. First, the shift reflects and deepens existing forms of anthropocentrism. The new perception articulated by Weetjens is not anthropocentric in the sense that it recognizes that nonhuman forms of perception exist, matter, and compose the world differently. However, these sensibilities exist to find exploitable modes of experience that harnesses the rat sensorium in the service of actualizing a providential vision of human life. Anthropocentric reason cultivates a broadening of perception to better engineer the world for human flourishing by capturing the world, the *Umwelt,* of rats. Here articulations of animal welfare depoliticize the process of capture and normalize a kind of servitude that forms the basis for rat demining and infectious disease identification. Second, in contrast, the perceptual shift might suggest a trend toward a form of multispecies justice that participates in what William E. Connolly calls a "new politics of swarming" in response to political violence.[93] Although helping concrete human communities remains a central goal of APOPO, the reconstitution of sensation that creates the possibility of rat demining also generates greater sensitivity for ecological well-being, coexistence with nonhuman others, and, as DeAngelo argues, possibilities for humanitarianism beyond a militarist ethos.[94] Thus, though in its infancy, this shift raises the question of whether humanitarianism's focus on human development and security is separable from the articulation of a broader vision of ecological interrelation facing dire conditions of agrologistics, colonialism, homogenization, and ecocide.[95]

This shift at the level of sensation may also explain why APOPO began to explore other potential humanitarian applications of rat–human technical assemblages. Again, even the most crude, anthropocentric application of rats as technology presupposes a mode of ecological encounter that remains unsettling relative to normative images of human experience. More than demining, the practice of TB identification showcases this shift in intelligibility and multispecies work. APOPO trains rats to identify TB using similar techniques to land mine detection. After a period of initial rat–human socialization and habituation, trials began using clicker training methods to help rats differentiate TB-positive samples. Using positive reinforcement techniques, rats began to distinguish TB-positive and TB-negative samples. Relative to the main standard procedures for TB detection in its countries of operation (smear microscopy), APOPO claims that HeroRATs increase the number of successful TB detections for its partner clinics by about 40 percent.[96] Rats can move through dozens of samples in a short window, expediting a task that would take lab workers days.[97] Research studies show that most of the trained rats are successful at identifying samples more than 70 percent of the time, with a relatively small false-positive rate. These results have been independently reproduced, tested in pediatric populations, and, in conjunction with other verification methods, used to improve disease identification in multiple places.[98]

But what are the rats doing that enables them to sense TB? Subsequent research indicates that TB produces particular "volatile organic compound [VOC] patterns."[99] These were "found to be distinct from those emitted from cultures of non-tuberculosis mycobacteria species or other bacteria that look similar under the microscope."[100] Humans also interact with VOCs when they smell decay, candle scents, and innumerable other odors, but, unlike rats, they have a far more limited propensity to differentiate these sources. The VOCs from TB are not only much fainter but also qualitatively distinctive in at least two ways. First, it is not the presence of the organic compounds themselves that is distinctive but their *pattern of interaction*. Put differently, many of the VOCs in TB are common to multiple species of bacteria, but the emission of these gases distinguishes the type of bacteria. Thus, "while no single compound was alone predictive of TB, a *bouquet* of VOCs in varying composition, quantity, and overlap were found."[101] This bouquet includes potential differences such as quantity of specific compounds, rates of emission, particular isomers, or their mode of interaction. Second, the pattern identification is a product of the rats' attunement to variations in the emissions of bacterial reproduction. What the rats are smelling for is not a single identifying feature of TB at an organic level, something akin to a bacterial essence, but a dynamic expression that occurs as TB cells interact with and territorialize their

HeroRAT Julius sniffs for TB samples. Copyright APOPO.

own environment. In the language of Deleuze and Guattari, rats have learned to sense, appreciate, and distinguish the "refrain" of TB in the flow of smells in a biochemical phylum. Humans also scent these patterns at specific moments, such as the odor of fermenting yeast, but most human olfaction is not attuned to the intensive differences that generate these odors. The irony here is that while diseases are often depicted as gaseous, such as miasma or the fumes of decay, most humans evolved to be texturally insensate to an ebbing, oozing, and throbbing of olfactory variation. Rats, despite being classified biologically as large mammals, richly tap into these microbacterial exchanges. As Stefanie Fishel argues, it is a form of antimicrobial political bias that causes humans to assume they are the authors of their world when microbial processes, symbiogenesis, fermentation, viral infection, and horizontal genetic transfer create the backdrop for so much of mammalian life.[102] In a sense, rats are more richly, more actively attuned to ongoing, subterranean interactions of microbial communities that produce ecological conditions.

The capacity to become an effective agent of infectious disease prevention and a laborer in the context of humanitarianism emerges from ecological interactions, but ones that differ from the version of demining dogs. Although the organic structures of rats and dogs share similarities (as with human animals)—as Manuel DeLanda might say, they all appear as actualizations of a topology of the abstract animal[103]—rats have a different degree of power for accessing this microbial plane. Though this capacity alone does not define the rat as a humanitarian, it illustrates how the types of labor defined as humanitarian build from disparate modes of sensation that collaborate and interact across distinct layers and scales of ecological life. Each of these complex aesthetic encounters, which involve organic and inorganic entities, is a contingently evolved, historical capacity forged though interaction rather than predetermined by the rat's biology or essence. Furthermore, the rats "become" humanitarian through a set of technologies that capture, study, train, and repurpose their ability to sense a different biological spectrum but also through joint, reciprocal, generous interaction with human partners. Yet, even in what appears to be the homeostasis of the laboratory, rats note intensities in the movement of the bacterial phylum and can link these intensities with pleasure. Tracking bacteria may seem dull, but for rats, it *appears* to be a source of intrigue, one that they return to with repetition. Using this potential is what enables humans and rats to interact such that humans exert a measure of control, but the rats' sensation is also not reducible to this means of control.

By experiencing the aesthetic contact with the bacterial spectrum, the rats' sensation and subsequent motor movement supply crucial information for the treatment of humans. In this way, human interaction with rat sensation is deeply ambivalent. Although the rats are given food rewards in exchange for their labor, giving these rewards produces an awkward ethical dynamic in the context of humanitarian action because the latter is premised on the welfare and rights of living humans. Though there are many tensions within humanitarianism, open exploitation of a coveted relationship to caregivers is hard to sustain even in a world steeped in anthropocentric privilege, especially if a rat's capacity for disease identification or explosive detection occurs because of a simultaneous shift in perception that entails being more generous or open to the capacities and experiences of nonhuman animals. As Vinciane Despret contends, "learning how to address the creatures being studied is not the result of scientific theoretical understanding, it is the condition of this understanding."[104] In short, APOPO is still in the process of working out how to articulate the relationship between human and rat; it is a moment of interspecies learning that begins with care as an apprehension of the rat's world but then resorts to the most general of understandings to supplement the inability to think across complex ecological difference, to understand the rats' gift other than through the lens of providential humanism.

Jacques Derrida's famous account of the gift offers useful insight on this point. Derrida argues that an aporia defines the practice of gifting. For Derrida, giving a gift creates an impossible burden or debt on the other that is ceaselessly annulled through a reciprocal response, even if it takes the form of a mere "thank you."[105] The genuine giving of a gift is therefore a constitutive impossibility because the moment a gift is given, the gift becomes part of an exchange that annuls the gift, but, Derrida contends, it is this very impossibility that necessitates giving. If a gift entails obligation, then it is not a gift, because it obliges a response; it is a simple exchange and contrary to the conditions of something given without return. This typifies most forms of "giving" in everyday interactions. Gifts are reciprocated, responded to with politeness as a mode of decorum to reestablish parity between the parties.[106] Humanitarianism also hinges on this idea of the gift, a relationship between self and other where the gratitude of the needy is a sign exchanged in response to the material support and investment of the powerful. Derrida concludes that the only possibility of a gift occurs through a kind of incommensurability, when the giver gives without realizing that what occurred occurred as a gift.[107] In this way, the gift takes place unilaterally and without an economy of exchange. Such a response is impossible because the very recognition of the gift destroys the gift as such.

This process characterizes the rats' engagement with TB and institutes APOPO's expression of indebtedness to a new subject that gives without return—the HeroRAT. Although rats might "receive" food, comfort, health care, and other items of biopolitical concern, it is an open question whether rats (or each rat in its individuality) understand these rewards in terms of an exchange. Rather, when a rat pursues the emissions of TB or the scent of the mine, it is discerning an intensity, and this discernment catalyzes land mine detection and the accuracy of medical diagnoses. If a gift occurs between rats and humans, then either it only happens on the condition that there is no realization that the act was a gift at all or it implies that rats and humans participate in a form of metacommunication in which exchange occurs across incommensurate species difference. In this way, even conservative interpretations of rats paradoxically make them into gift givers par excellence because they do not participate in the human, all-too-human symbolic economy around the gift, or one is forced to admit that rats and humans are capable of contestation that enables critique of humanitarian politics. Put differently, the ecological attunement of rats, linked to a human assembly, becomes the condition of possibility for helping to treat people affected by TB, but at the same time, this very ecological attunement makes the rats unaware of the existence of the gift at all. From one angle, the rats are an anchor, the source for the material capacity to deliver aid and, as such, a vital condition of possibility for ethical humanitarian action. From the other angle, this ethics hinges on what Derrida might refer to as the *disparate* or the incommensurability of the self and other.[108] APOPO's ethical response, pushing the rats to the forefront, occurs because the organization functions as an apparatus of capture, converting the rat's sensibility into a means of labor and entering rats into an economy of symbolic and emotional exchange. In doing so, the gift's impossibility is obscured but nevertheless haunts the exchange in the form of the livelihood of the rats. APOPO's response is to center rats, celebrate rats, and care for rats to address what would otherwise come across as a problematic indebtedness. There are thus two modes of humanitarian ethics involved here. One is an economy premised on an incommensurable exchange that Derrida identifies as an impossible form of giving. In this mode of giving, it is the rats' difference that grounds the incommensurable relation that makes a gift possible. In this case, the otherness of the rats' (and other nonhuman others') interactions might constitute the only conceivable gift and, in Derrida's sense, an originary anthropocentric erasure that grounds the possibility of a symbolic economy. The other is a symbolic transaction in which the gift becomes a form of exchange. What is strange is that, in this case, it is the rat that ultimately provides the impossible gift, whereas the humanitarian organization, which focuses so much on relations of debt, converts this gift into an affectively imbued, highly sympathetic form of exchange. As such, the ecological contact of the rat becomes the basis for both extending life and reinforcing forms of humanitarian power.

Derrida's reading illuminates how the insufficiency of existing discourses on ethics confronts the problem of incommensurable differences. The gift that emerges from the rats' interaction with chemical vapors could be appropriated without any recognition or acknowledgment, a mode of reduction that reduces rats to nothing more than an instrument of anthropocentric reason. Yet, the stretching of perception involved in multispecies connections, the forms of curiosity and experimentation that characterize APOPO's organizational self-articulation, also involve an aspiration to remain aware of the impact of a set of actions on otherness and a fascination with the ecological sensitivities of rats. By constructing the mission of rat mine detection or disease identification on these premises, APOPO admits rats into a realm of objects of care, where care is understood not simply as empathizing but as concern for the existence of something strange, something nonhuman.[109] In contrast, as with other types of nonhuman humanitarianism, rats could simply be written out of any symbolic exchange, removed from an economy of the gift altogether. Here, instead, the rat is positioned as a humanitarian in a way that sets up an economy in which incommensurate sensations facilitate a communication that can never be fully reciprocated, in which the human world of gifting cannot be successfully extended if it continues to rely solely on anthropocentric terms. As a result of this process of stretching sensitivity and encounter, it becomes more difficult for APOPO to exclude the rats from humanitarian considerations and, indeed, nudges it toward countering the hostile myths that develop from agrologistic prejudices. However, the symbolic economy APOPO embraces also does not have the capacity to reconcile with the rats' gifts. Its foundation in anthropomorphism ensures that there is no bridge across intensities of rat and human experience, much as both rats and humans live, work, and communicate with one another through their differences. Moreover, APOPO's understanding of the economy of giving inherits many norms and ideals from classical paradigms of humanitarianism. HeroRATs exist in a liminal space pulled between two different possibilities: on one hand, the attractors of dominant emotions, affects, discourse, and values associated with humanitarianism and the strictures they seek to impose on both the category of the human and the understanding of political relations and, on the other, the strangeness of rats scenting the emissions of bacteria and chemistry, the differences and connections between rats and their human interlocutors. The gift of rats thus becomes coupled to a dominant set of exchanges that work to convert rats into champions of humanitarianism, linking up with hegemonic tropes about humanity but also unable to shirk a disquieting sense that this is insufficient to the rats' gifts and the weird encounters from which they arise.

MULTISPECIES GIFTS, RATTY JUSTICE

The problem of humanitarian law, violence, and the possibility of justice directly pertains to the context of HeroRATs. By prioritizing rats, caring for their welfare, centering the possibility of humanitarian work in their ecological sensitivities, new possibilities for humanitarian action emerge. At the same time, APOPO describes rats as doing good work for vulnerable communities and, in making this claim, turns rats into celebrity figures of nonhuman humanitarianism. This model reproduces across species difference a recurrent theme in humanitarian literature in which "humanity [was] regarded as the hero of a narrative in which liberty continually expands."[110] If a flicker of justice appears in the new articulation of this theme, it occurs in the double move toward a kind of equity or care that calls into question the boundaries between human and nonhuman, by stretching interpretations about how care emerges as a result of different sensitivities while simultaneously articulating this interest in the hope of an expanded humanitarian legal order.

The rats give in another way. They do not contribute to humanitarianism solely by supplying information that is useful for subsequent human intervention on an explosive ecology or a diseased body. It would also be absurd to claim, without hesitation, that rats have an awareness of the scale of humanitarian projects. Hence the rats contribute to this process, but with an incomplete knowledge of the gift. But how does this differ from human experience in which abstractions, rather than perfect knowledge, define relations to myriad complex logistical networks and singular contexts of suffering and care? These are, ultimately, fantasies that permit exchanges and the symbolic process of gifting. Consequently, humanitarian ethics are not a result of a particular distribution of knowledge, nor reducible to awareness of the predicament of the other, but subsist from the genres, sensations, and scripts when they encounter divergent entities and situations. There is a danger in treating the supposed ignorance of rats as the basis of an ideal gift because this position turns incapacity and subordination into saving grace. Worse still, this approach misinterprets what is occurring in the case of HeroRATs. The epistemological limitations of rats with respect to human affairs are not a symptom of the rats' inferiority (cognitive, emotional, or otherwise) but a gap between disparate modes of aesthetic engagement with politics and ecology. The incommensurability of human-built networks (which are themselves obviously much more than human) with rat olfaction signifies that those two heterogeneous systems of interaction are just starting to enter into communication with one another. Because human systems inevitability anthropomorphize or "humanize" this interaction, they reach a limit in their ability to articulate the possibility of the rats' contributions without mooring them in terms of human life and well-being. The rats value remains ineffable in relation to these systems of exchange. Mawaga's award, which was briefly discussed at the beginning of this chapter, illustrates the friction of grafting an economy of human gifting that already struggles to articulate the relationship between gift and debt across an incommensurable divide, to come to terms with processes of desire and enjoyment that do not coincide with human terms.

In his reading of the mystical foundations of the law, Derrida draws a distinction between legal authority founded by right, which, because of its explicit construction, is prone to critique, and justice, which he contends is undeconstructible. As Derrida puts it, "it is the reconstructible structure of law . . . that ensures the possibility of deconstruction. Justice in itself, if such a thing exists, outside or beyond law, is not deconstructible. No more than deconstruction itself, if such a thing exists. Deconstruction is justice."[111] Derrida's argument is the subject of a vast commentary often focused exclusively on debating the intricacies of his thought.[112] Derrida's position is that law remains subject to deconstruction and reconstruction because of the arbitrariness of its formation. Legal reforms, the overturning of a precedent, and even the possibility of revolution persist in this plasticity at the heart of law. But none of these constructions coincide with justice. Rather, justice remains an impossible but necessary supplement to any legal regime. The idea of justice is beyond their scope or powers of comprehension, but it is also toward the possibility of justice that every legal regime extends, only to have the concept of justice recede.[113] Justice exists in an interstice between its invocation and actualization, its performance, and its difference from this performance, always, to come. For Derrida, the constitutive impossibility of justice grounds the possibility for reform (and here reform does not mean a modest change to the legal order but a reconstitution of the form of law) beyond any specific reconstruction of law. Deconstruction and justice both exist because of the structure of law itself.

At first glance, humanitarianism may not appear to be a problem of legal order. It seems to reject as foundationless any illegitimate violence against an individual human, but in doing so, it also tacitly affirms the importance of legitimate violence as defined by state and other authorities—hence the distinction between humanitarianism and nonviolence.[114] In practicality, there are few humanitarian laws and mechanisms of enforcement for them even fewer in a geopolitical era characterized by the machinations of power politics.[115] Nevertheless, humanitarianism functions as a legal regime in Derrida's sense because it invokes a set of constructible and deconstructible boundaries in the articulation of its most basic distinctions—human–animal, individual–community, universal–particular—as it draws from a lineage of political theology grown from the agricultural and dedicated to a specific conception of the good under the aegis of embedded cultural norms.[116] To put it differently, humanitarianism is an aspiration for a different set of rules contaminated with new metaphysical prejudices. These aspirations, to varying degrees, acquire the force to be enforced. Humanitarianism constitutes a type of law, albeit one very different from the group of regulations and declarations typically understood as a legal code. It desires legalistic power even if its enforcement is wanting.[117] If humanitarianism also breaks from dominant regimes of legal power and exists in a sometimes contorted proximity to justice, this is because it makes a claim against one conception or model of discriminatory state power, only to replace it, as Talal Asad argues, with another "metaphysical conception of life" that justifies its own violences.[118] Humanitarianism provides a new construction of legal order, albeit one with a different distribution of equities and inequities.

Read in this way, justice in humanitarianism occurs in the form of an impossible deferral because the most robust, most universal claims about the equitable and emancipatory treatment of humans merely strive toward, never to arrive at an instantiation of justice. For Derrida, all of this is quite understandable. Law's emergence is anchored to divisions regarding life and death, presence and absence, self and other.[119] In this sense, law is intimately biopolitical and necropolitical, determining what forms of difference, life and nonlife, fit within the confines of a legal order. The underside of the promotion of the universal is the creation of a subtle but discriminatory politics. What does it mean to be human? What constitutes a right? What types of inhumanity or nonhumanity are palatable? These are foundational problems for humanitarianism as a legal order because they reveal the arbitrary conventions that constitute it. As the introduction argued, the appeal of humanitarianism as a frame for understanding politics is partly a result of the way it helps to tie together the loose ends in these ceaseless divisions between human and nonhuman, making humane action, emotion, and reason into a self-referential knot in which the nonhuman otherness appears as a limit. Justice, following Derrida, vanishes in the insistence of this moral paradigm because this paradigm erases nonhuman lives and deaths.

Yet, if Derrida is correct, it is in the difference marked by this divide that the possibility of justice subsists. This model of an impossible justice reiterates the structure of an impossible gift. In the context of rat humanitarianism, the incommensurability of difference between rats and humans means that the rats' capacity to scent cannot be successfully articulated by symbolic economies or legal aspirations. At the same time, the point of contact between rats, humans, and land mines and microbes, regardless of the symbolic gesticulations, ricochets through explosive ecologies and TB-positive patients, changing lives and landscapes. Rats, in this regard, introduce a break, a discontinuity, within humanitarianism as a nonhuman that gives without knowledge of the symbolic economy of the gift and, because of this encounter, creates new possibilities for life. An incommensurable otherness founds the potential for a more expansive practice of generosity, but also an encounter, a zone of metacommunication, in which subsists the crack to consider a different invocation of justice.

The rats are quasi-miraculous from the perspective of humanitarian symbolic economies. They ask nothing in return for giving the gift of labor and life and intervening in problems that create death and suffering for human communities that humans can never fully articulate or convey. Normative humanitarian ethical systems thus cannot fully account for the rats' gifts, so they default to the anthropocentric values and discourses they can articulate, revealing the insufficiency or the difference that marks human models of concern, while pointing to a different, if ineffable, possibility for justice. The fact that humanitarian institutions oscillate between governing rats as a species, genuflection, and open expressions of gratitude demonstrates how the rats constitute an aporia in the principles of humanitarianism. Historically, rats are the scourge of Eurocentric humanitarian nightmares as they problematize agrologistic principles and spread infection. Here, however, the rats become the source of flourishing for humans. This flourishing creates a dilemma for existing discourses of recognition and gratitude, because the gaps between humans and rats, the affective disgust with rats as bringers of plague, and the impressive capacity of rats to detect explosives and disease set limits on human capacities ever to express thanks using anthropocentric terms, in effect formally annulling the possibility of reciprocating a debt or responding to the gift in advance.

There are, quite literally, no words that adequately communicate thanks for the gifts of rats. It is the ecological weirdness of rats, their interpretation of bacterial odors and chemical signatures, that is the source of a gift, an ecological difference that is a source of frustration for agrologistics because it enables rats to become parasites to the everyday function of consumption and symbolic exchange, much as it also saves lives. Here the rats open up a glimpse of the tensions that humanitarianism confronts as it mutates, because they expose its biopolitical character, its division and governance of life, its symbolic exchanges of gratitude and power, and the way care often functions on relations of parched debt and reciprocity. Instead, care subsists in the possibility of assemblages built around incommensurable ecological encounters never perfectly captured by the anthropocentric categories that articulate it. Humanitarian inter-action abounds with these dissonances. It subsumes rats in a biopolitics grounded in anthropocentric reason, a living tool, used for the extension of human flourishing, a process that converts the nonhuman villains of agrologistics into a resource for the extension of this very paradigm of power. It encounters a patchwork of strange sensibilities in which the principles of human life do not define flourishing or ethics but instead learn to live, finally, with the gaps, hiccups, and impossibility of seamlessly coexisting with others, whether human or nonhuman, and deploys these as a form of ethical curiosity and symbolic exchange. If, in one sense, HeroRATs were made through a practice of capture, observation, disci-pline, breeding, socialization, and behaviorism, in short, control at the level of the reproduction of their form of life, then these rats also helped to open a laboratory for the articulation of new models of humanitarian ethics, grounded in mutual care, that remain open to nonhumans. The oddness of a gift that cannot be a gift may transform into a familiar hu-manitarian refrain with furry nose, but it also entails, simultaneously, an exploration of multispecies justice in which rats aid humans and humans contest violence against nonhumans.

Thinking through this ethical predicament, in contrast, entails more than simply caring about the lives of rats as they perform labor for humans. It requires humanitarianism to rethink care not as an abstract or normative ideal, nor as a simple reciprocal extension of feeling to nonhumans, nor even as an extension of the limits of human perception, but as a vexed, partially hidden set of relations in a process of becoming. Rather than establishing ethics based on the minimal threshold of being "merely" human or "resembling humans" or "doing work for humans," it involves committing to a foray into otherness, to the possibility that something nonhuman can generate previously unknown forms of care. Indeed, if the rat's labor using its ecological difference constitutes a gift that is not reducible to existing humanitarian discourses on symbolic or sympathetic value, then it signifies that possibilities of care, generosity, and engagement celebrated by humanitarianism do not originate with the human at all but with the unknown encounters proffered by contingent interactions.

3

The Gift of Milk and
the Contingency of Hunger

In one of its videos, Heifer International (HI) poses the question, "Have you ever wondered what life as a baby goat looks like?"[1] This inquiry is followed by a series of images taken from a camera mounted to the head of a nameless juvenile goat. The camera tracks the goat's movement around different areas of a yard, showing short encounters with other goats, chickens, and a pig. The viewer also sees wire fencing, a trough, a barn, and a small feedlot. The height of the camera and the jerky motion of the video assure the viewer that this is, indeed, a small goat's view of the world. Paired with music, the promotional emphasizes the charm and frivolity but also symbiotic aspects of the infant goat's life and links these indirectly to the possibility of addressing chronic hunger.

"Have you ever wondered what life as a baby goat looks like?" resonates with an entire genre of memes and videos that use endearing images of nonhuman animals for entertainment. The video's prompt, to question what the world looks like from the perspective of the other, also echoes a theme common in humanitarian literatures where the audience is asked to place themselves in the position of a fellow human in need, but here, the other is not a human but a goat. Like the promotional materials of many humanitarian organizations, this framework positions the viewer in a place of power, externally observing the life (and more often the suffering) of a typically racialized, classed, and gendered other who, in turn, acknowledges the benefits they receive and dignity they feel as a recipient of humanitarian compassion.[2] Even in the most uplifting, positive, asset-oriented versions, the presence of the camera structures the exchange being celebrated or promoted.

However, by shifting the locus to the life of a goat, HI signals, at minimum, a new riff and possibly a break with the traditional organization of this genre. The video shows the life and labor of the goat as part of a multispecies world, and the goat on display may resist or, at least, not live a life entirely compatible with humanitarian genres. First, the movement is jarring, requiring the producer to synthesize short clips of the goat's daily life to smooth over the clunky disruption of the visual spectrum as it juts, bobs, and weaves with the goat's desire to smell or gnaw on what is underfoot. The horizon, typically established by a strong distinction between foreground and background, is hard to identify in the clip because of the sudden proximity to the earth, the swift appearance of objects in the goat's view, and the goat's interest in things deemed mundane by a human audience primed for excitable visual stimulation. While the video's rhetoric captures the potential joy of being a baby goat, the world this visual rhetoric implies also is, at times, incompatible with the nonhuman interests of the baby goat.[3] Life, from the baby goat's perspective, may be displayed but is not fully captured by humanitarian visual genres.

This chapter examines the appearance of caprines and bovines, more commonly called goats and cows, in humanitarian relief efforts. Though the use of domesticated goats and cows in agricultural practice is millennia old, their participation in practices of humanitarian aid is much more recent. In this context, the chapter argues that this type of humanitarianism effectively relies on the labor (and the lives) of cows, goats, and chickens and, furthermore, that, much like the baby goat, these nonhuman animals transform both the aspirations and practices of humanitarianism. Goats and cows have lived as part of human communities for thousands of years, and their bodies bear the markings of evolutionary changes brought about through this coexistence.[4] The labor of goats and cows in agricultural settings is thus nothing new, but humanitarian deployments of these animals is neither a return to an ancient farming practice nor a nostalgic reproduction of a bucolic image of happy premodern life. Rather, it is a contemporary strategy for addressing food insecurity created by conditions of deprivation and subsistence. In one sense, goats and cows are instruments, framed by anthropocentric reason, that facilitate economic development, ecosystem management, food production, and emotional well-being. As with HeroRATs, the assumptions of agrologistics play a powerful role in defining humanitarian ethical relations with goats and cows. Reading humanitarian action through the prism of the activity of goats and cows helps illuminate these dimensions of food aid as well as the normative image of human life that informs humanitarian governance.

The chapter begins by outlining the turn to goats and cows as instruments of humanitarian aid during the second half of the twentieth century and contrasts this practice with the longer human coexistence with bovines and caprines. Although dairy farming has unique economic benefits, in its contemporary form, this practice presupposes a particular image of human–nonhuman relations to explain how nonhumans contribute to humanitarian efforts. The second section examines goats and cows through an ecological lens. What capacities of these mammals make them amenable to use by humanitarian organizations? How do cows and goats engage diverse ecologies and enable changes in agricultural and economic production? Answering these questions helps to explain what capacities make cows and goats ideal humanitarian laborers. The third section analyzes milk as one of the main products of goat and cow labor. It begins by unpacking the presentation of milk as a gift, positioning nonhuman animals within a moral economy as providers of humanitarian benefit. It then turns to the cultural and dietary assumptions surrounding milk consumption. As numerous studies now document, dairy products are not geographically or culturally neutral but laced with unmarked, frequently racialized expectations about what constitutes a healthy body and viable body politic. The fourth section builds on this point by discussing how humanitarian organizations understand the problem of subsistence. Here the chapter argues that assumptions about the good life, contingency, and other metaphysical anxieties structure the delivery of humanitarian services, which focus on a particular form of life, alongside adequate nutrition, as a component of the underlying problem. Framing this relationship in terms of agrologistics demonstrates how contemporary forms of power structure the articulation of hierarchy between humans and nonhuman animals by articulating a fear of specific kinds of contingency. The chapter explores how, paradoxically, goats and cows help to rescue humans from this condition not only by providing valuable food and economic resources but also by reinforcing anthropocentric norms. In short, agriculture becomes a remedy for a set of contemporary existential anxieties.

The final section addresses the role of meat production in humanitarian farming. It argues that, unlike rats and dogs, goats and cows are not framed as active participants in gifting or as full-fledged humanitarian actors because of the paradoxes this would engender in relation to practices of nonhuman animal slaughter. Indeed, if goats and cows were understood as humanitarians, then their consumption would place humanitarian organizations in a perilous position of recommending a practice akin to cannibalism. This arena of nonhuman humanitarianism is marked by tensions as it wrestles with the problem of recognizing and addressing violence and deprivation, seeking to articulate a new model of animal welfare that breaks with dominant practices of industrial agriculture and the production of meat as part of its humanitarian service. In contrast to HeroRATs, which are caught in a symbolic economy of gifting that cannot fully articulate their contributions, goats and cows are excluded from the discourse of humanitarian laborer, even as milk, a result of their biological labor, species-being, and ecological attunement, constitutes the main product of these forms of humanitarian aid, but nonetheless their well-being remains a matter of humanitarian concern. The chapter concludes by reframing subsistence as well as human–nonhuman collaboration as harboring forms of resistance to these paradigms and providing further clues to the forms of care that reside within humanitarian practice.

THE GOAT OF HUMANITARIANISM

Nonhuman animals have always contributed to humanitarian efforts in the form of labor, milk, eggs, and other foodstuffs. However, goats, cows, and other domesticated nonhuman animals associated with agriculture became formal objects of humanitarian aid during the middle of the twentieth century through the interventions of HI. HI's model of humanitarianism is distinctive because it provides nonhuman animals, rather than foodstuffs, as the substance of aid. This approach to hunger relief was developed by Dan West, a member of the Church of the Brethren, a peace activist, and founder of HI, who came up with the idea as an aid worker during the Spanish Civil War.[5] West observed that humanitarian relief efforts depended on a stable flow of supplies, which was inadequate in cases of severe famine, extreme poverty, or mass displacement. Upon returning from his service, West proposed a practice of humanitarianism that would deliver dairy cattle to those in need. By providing cattle directly, West thought HI could foster a sustainable form of dairy production that would alleviate chronic hunger, thereby eliminating the contingencies associated with imported foodstuffs, especially in emergency situations.[6] Cows were not just a means to feed someone for a day but a mechanism of providing families with the ability to create a livelihood and support themselves.

What began as a small-scale, faith-inspired effort slowly evolved into a complex, transnational, multispecies endeavor. Formally established as the Heifer Project (HP) in 1942, and later incorporated in 1953, the nascent organization made its first delivery of livestock in 1944. During

Dan West distributes clothing to Spanish women and children during the Spanish Civil War in 1937. Copyright Church of the Brethren/Kermon Thomasson.

its first two decades, HP primarily sent cattle to postwar Europe and later to Puerto Rico, Mexico, and the southern United States.[7] Eventually, HP expanded its focus to include exporting calves and providing expertise to communities in Asia, South America, and Africa.[8] HP also moved beyond dairy cattle to include goats, sheep, pigs, poultry, rabbits, water buffalo, and bees in its deliveries. Over the course of seven decades, HP, later renamed Heifer International, provided support to approximately thirty-five million families through tens of thousands of animal deliveries.[9] HI also transitioned away from delivering animals to purchases in local or regional markets. As its operations became more complex, HI expanded its services to include other forms of direct aid, such as strategic planning advice, training in animal husbandry and farming practices, climate change adaptation resources, and combating gender inequality.[10] Nonetheless, the benefits of farming with nonhumans remain a signature part of HI's approach to humanitarianism and stimulating economic development.

The empirical evidence that nonhumans improve agricultural and economic outcomes is strong, and West's instinct that humanitarianism's crisis orientation meant that it was ineffectual against both chronic and acute disruption to global food production bears out.[11] By creating cycles of dependence, this traditional model of humanitarian relief is also quite susceptible to cooptation by other interests.[12] In contrast, HI maintains that small-scale farming provides multiple benefits to developing communities by catalyzing agricultural production and local economic transaction. Moreover, HI's practice builds from the human–nonhuman interaction to consider broader ecological and economic questions about the effect of different modalities of agricultural production and environmental use. The organization shifted from a simple focus on animal deliveries to a more complex concern for security and well-being in relation to climate change and to greater consideration of how small-scale coexistence with nonhuman animals strengthens local ecologies. Nonhuman animal labor is a crucial, transformative ingredient that enables this process. When mammals produce milk or are butchered for meat, they convert indigestible plants, which are of little caloric or nutritional value to humans, into consumable proteins and calories for human communities. This capacity constitutes what Nicole Shukin might call "animal capital" and enables the creation of commodities of milk, eggs, meat, and other goods that generate sustainable economic activity and allow local communities to thrive without pressures of global agricultural markets.[13] Nonhuman animals also contribute their labor by clearing plots for harvesting cereals and other crops, performing different feats of strength and endurance like plowing fields, and generating manure that fertilizes crop production. These various products, linked with human labor, jump-start agricultural and economic investments, which, in turn, slowly redress elements of chronic hunger and underdevelopment.[14]

For HI, hunger and development point to an underlying problem, subsistence, that places communities at unique risk of food insecurity: "there is a lot of uncertainty that comes with [subsistence] farming. Issues like climate change, lack of technology, illiteracy, and access to opportunities pose a great threat to [farmers'] source of income."[15] Sending nonhuman animals to developing communities accelerates the movement beyond subsistence. As one publication puts it, "while subsistence provides goods and nutrition for the time being, an active marketplace creates incomes, savings, education, employment and opportunities for communities to retain their people, develop new skills and increase earning potential."[16] The transfer of nonhuman animals forms a key part of this project and sets the stage for a broader transformation of civic and political relations alongside economic development. These opportunities enhance the resilience of subsistence communities to the forces of globally integrated markets and regional crises, anchored in the assumption that farming with nonhumans reduces food insecurity, a subtle shift from HP's earlier faith-inspired model of humanitarianism.[17]

As part of its practice, HI created a set of principles to guide its vision of humanitarian work. One of its most important principles, foundational to many of HI's other stated beliefs, is a commitment to "passing the gift."[18] This principle asks the recipients of HI's assistance to commit to passing on some of their animals, supplies, or knowledge to their neighbors. The commitment is designed to incite the development of local markets and connections while providing for regional autonomy, independent of external humanitarian oversight. As HI argues, milk and other nonhuman animal–sourced foods are "the fastest way to get vital nutrients to people that need them most" but also a "farmer's . . . most valuable assets."[19] In an economy of giving, farmers are positioned as both recipients and givers of gifts. These gifts and exchanges sprout new opportunities; new gifts, including "a network of partners, suppliers, and even customers"; chances for "inclusive economies . . . to grow organically"; the opportunity for "sharing lessons" on business and farming; and "access to new markets," ending gender inequality, empowering "community animal health workers and feed suppliers," and "defending the planet" from climate change and environmental destruction.[20] In short, once enacted, the principle of passing on the gift generates a series of commitments and network effects that follow flows of milk through a mesh of social and economic relations to help create durable, resilient communal bonds. In much of this literature, HI frames the cows and goats involved in this process as what Bruno Latour might call an "intermediary," or a set of actors that are a part of a social network but do not introduce significant changes to it.[21] While HI articulates an explicit commitment to animal welfare, ecological sustainability, health, and nutrition, goats and cows are "valuable assets." Proper care is framed as important to "producing high-quality milk" and other goods.[22] Animal health and proper training are meant to keep "livestock in top conditions."[23] HI interprets goats and cows through the prism of anthropocentric reason even as it also takes into consideration their well-being.

HI also embraces the emotional aspects of farm animals. Goats and cows are a wellspring of moving images, affective and caring relations, and semiotic content. While the delivery or the purchase of animals is now a more widespread humanitarian practice, human–nonhuman companionship and interspecies cohabitation incite fantasies about the virtues of humanitarian labor.[24] Like the promotional video described in the introduction to this chapter, this rhetoric distinguishes the significance of nonhuman labor in implicit contrast to the efforts of other models of humanitarianism. Like the cases of rats and dogs, anthropocentric feeling, shared human–nonhuman contact, paradoxically humanizes humanitarian enterprises. Thus, for example, HI's goats, cows, pigs, and chickens are frequently described in the terms of economic development and agricultural practice but also consistently depicted as "companion species," making bread, or in this case, milk, together with humans. This discourse suffuses anthropocentric feeling with humanitarian affect by connecting the recipients of aid with a cuddly, nonhuman farming companion.[25] HI, in particular, advances this imagery as a move against the deficit-centered narratives that often define humanitarian representation by celebrating the positive energies that emerge from linking people with new opportunities provided by nonhuman animals. The imagery works because, as Andrea Laurent-Simpson documents, there have been major demographic shifts in the Global North over the past five decades to embrace the family as a multispecies endeavor.[26] Nonhuman animals, primarily cats and dogs, but also other furry, social mammals, have progressively adopted novel roles and identities in family structure, and families have diversified the species they make home with. Bonding through "animal affects" is a strong way of activating the set of intimacies and sympathies that promote donation and one of the key elements of HI's persuasive rhetoric. This practice uses nonhuman animals to reinforce a specific model of family bonding. In effect, the friendly, charming, evocative goat may inspire a greater emotive response than another classical form of humanitarian visual rhetoric, the suffering child. Deploying goats and cows in this way hinges on showing *implied connections* between nonhumans and their smiling human counterparts. The imagery creates a distinctive humanitarian rhetoric that defines itself by displaying nonhuman animals as a brand that facilitates easy linkages between a farm, agricultural life, dairy production, family life, and full bellies.

Goats and cows are part of the visual politics of humanitarianism just as they are hoofs on the ground, producers of milk, fertilizer, and meat. However, the content of this rhetoric, which emphasizes sociality and happiness, exists alongside practices of animal slaughter. Indeed, humanitarian organizations that deal with nonhumans are not shy about the fact that the nonhumans they provide are killed for the purposes of the sale and consumption of meat. Nonhuman companionship thus exists in a tense relationship with the possibility of violence against nonhumans. This duality is common, even constitutive of contemporary human–nonhuman animal relations at a global scale. As Cary Wolfe contends, nonhuman animals confront an odd condition, because they are biopolitically framed as recipients of both intensive care and oversight for the purpose of mass slaughter (in the case of chickens or pigs, living in industrial agriculture, for instance) and, simultaneously, unprecedented veterinary, emotional, psychiatric, and social support (in the case of domesticated cats and dogs).[27] Rather than understanding this tension as a contradiction, it forms a part of a much longer legacy of interpreting nonhuman animals as fundamentally content despite their exposure to cruelty, pain, and death.[28] Happy animals are a signature part of many forms of meat production, and the mutual existence of enjoyment and suffering occurs in other instances of animal cruelty.[29] In agrologistic terms, nonhuman labor that facilitates the extension of human life drives this process regardless of its quality. Hence dogs and cats, which offer emotional and mental support, receive one type of attention, whereas goats and cows, which also engage in affective labor but operate as agricultural capital, receive another. This tension impacts how humanitarian organizations, such as HI, view the labor and value of animals as participants in practice of passing the gift.

When Dan West initially developed a model of relief based on dairy farming, the focus was on the impact that cattle could have on areas that desperately needed sustainable food sources. Nonhuman animals were implicitly understood as a means to an end, an extension of farming practices from the mid-twentieth century American Midwest to Europe and, later, the Global South. Promoting human–nonhuman companionship was not as focal to American midwestern farming communities because, through their experience, they understood that agricultural work involved drudgery, fertilizer, muck, and arduous labor and that nonhumans were a means to an economic livelihood. The emphasis on nonhuman animals as affective participants in humanitarianism reflects a shift in the disposition or biopolitical framing of the relationship between farms and animals. This presentation of farm animals emphasizes their domesticated sociality partly for the purposes of producing an audience, turning an evolved bovine and caprine capacity for coexistence with humans into evidence of an enviable affective milieu.[30]

In offering a mechanism to address hunger and economic underdevelopment in the form of milk, meat, agricultural expertise, and economic resources, HI and other organizations tap into a shift in the culture of multispecies relations ongoing in the United States, Europe, and elsewhere, one that frames animals as an integral part of the milieu of successful communal life. A deep cultural cynicism may be detected in this shift because it presupposes that global economic injustice and malnutrition are not sufficient cause for action for audiences in the Global North, who, instead, respond more forcefully to the friendly aesthetics generated by human–animal sociality than to information about the scale of malnutrition and suffering. Nonhuman animals thus contribute to the emergence of optimism about humanitarian work, shifting the framing of humanitarianism, which often employs images of crisis that are tragic, apolitical, and constituted by misery.

In this sense, HI's reframing of dairy farming as a type of humanitarian work has made two changes to humanitarianism. First, it has introduced a new practice of humanitarian outreach centered on gifting nonhuman animals and assisting with the development of sustainable agriculture. Second, it has shifted the semiotics of humanitarianism by folding the lightheartedness of human–nonhuman companionship into humanitarian genres. Unlike land mine detection dogs or HeroRATs, whose material changes involve finding new capacities or new uses for existing nonhuman capacities, goats and cows have labored in agricultural production for millennia. Although technical improvements and the industrialization of agriculture made profound changes to these processes in terms of intensity, scale, and methods, the basic process of feeding domesticated bovines and caprines grasses and grains to produce milk and meat is anything but novel. Though often depicted as an act of unilateral force by humans against nonhuman others, growing evolutionary and archaeological evidence contests this one-sided explanation for the rise of domestication.[31] In contrast with contemporary, historically intensive forms of extraction and violence against nonhuman animals, some biologists argue that domestication of goats and cows resulted from at least some form of mutual benefit, including, for nonhuman animals, more consistent sources of feed, protection from nonhuman predators, favorable access to water, and sociable relations with humans, all providing hypothetical benefits to individual nonhuman animals and producing a viable environment for their lives and reproductive success. Furthermore, the discovery of

labor and of milk as a source of protein derived from nonhuman biological capacity may also be read as contributions to the "domestication" or
"domiciling" of some human communities as it prompted a transition
to build around more permanent multispecies agricultural settlement.[32]
Domesticated animals, including humans, display traces of evolutionary changes, such as transformations in morphology, that resulted from
the developmental plasticity these shifts involved. Tameness, a trait
often associated with nonhuman animals but also arguably descriptive
of many humans, accompanies morphological variations that constitute
the preconditions for making agricultural space.[33] Agriculture is less fully
human than is traditionally depicted and, while historically dependent on
continuous applications of unidirectional human power, is also a porous,
contested, multispecies assemblage. This fact should not, however, obscure the broader context of massive, anthropocentric violence against
nonhumans that occurs as part of agricultural practice, particularly in the
context of contemporary intensive factory farming. As Kathryn Gillespie
demonstrates, not only were cattle instrumental in the development of
early capitalism but the commodification of dairy cows (and, presumably,
other livestock) extends to include life, death, and even decay to extract
surplus value with deadly consequences.[34] Moreover, as Fahim Amir, Jason
Hribal, and others show, nonhuman agency, in the form of discontent,
resistance, and flight, plays a powerful role in the construction of agricultural assemblages.[35] As Sarat Colling argues, "the global destruction of
animals' habitats and the environment due to animal farming—prompt
the resistance of animals at increasing scales" and constitutes a key part
of struggles against the forms of violence fostered by hierarchical global
capitalism.[36] Read against this background, the delivery of goats and
cows builds on a long-standing practice of nonhuman domestication,
and adaptations among these nonhumans, but it also extends a relation of commodification common to capitalist relations to bovines and
caprines even as HI breaks with many practices dominant in industrial
agriculture. In this sense, HI's contributions concern the symbolic and
affective changes in humanitarian practice in the context of communities
that receive nonhuman animals, in the discursive justifications offered for

humanitarian action, and in reproducing dominant forms of agriculture as part of global capitalism.

ON THE MATTER OF MAMMALS AND MILK

Goats and cows became humanitarian laborers in part because they make milk. Although milk is taken for granted in groceries in the Global North, producing milk is a complex process that places many burdens on bovines and caprines.[37] Contemporary milk production involves multiple scales of governance, from the bodies of individual cows and goats to practices of livestock management to nutritional materials available in local ecology, infectious disease prevention, multispecies reproduction, and birth. Outlining this process requires a detour into the history of agricultural production, the role of grains and grazing, and feedback loops that condition cows, goats, and human farmers that set the stage for farming development as a humanitarian practice.

In the broadest sense, milk exists as an evolutionary adaptation of mammals to facilitate external growth of infant and juvenile offspring.[38] Milk is so central to the categorization of this set of nonhumans that the Latin term *mamma,* from which both the words *mammary* and *mammal* derive, means "breast," in reference to the capacity for making milk.[39] Mammalian bodies are adapted in many ways to enable milk production. Prototherian, metatherian, and eutherian mammals all secrete milk through slightly different mechanisms.[40] For instance, in prototherians, such as the platypus, both male and female sexed bodies develop functional mammary glands, whereas metatherians, such as possums, require the young to crawl into a body space called the marsupium to acquire milk. Each species and singular members of these species also have varying levels of fat, sugar, protein, micronutrients, and antibodies in their milk; different placements, lengths, and numbers of nipples (or other abdominal muscles responsible for milk release); and distinct habits or routines of milk production and consumption.[41] Some nonhumans consume milk for only a few days, whereas other species extend the production of milk for years. Milk production and consumption are also dynamic processes often requiring an infant animal to prompt its mother, father, surrogate, or caretaker to continue making milk in a feedback process after birth. The dynamism of milk production and consumption makes a quintessentially biocultural process, in Samantha Frost's sense of the term, as the habits of milk making and eating; the composition, taste, and nutritional value of milk; and even mechanical or species participation vary over time as an infant continues to feed.[42] All of these capacities are shaped by underlying ecological factors, such as available calories and nutrients, stress on the creatures in question, and other variables.

Humans are one of the few animals that, in some cases, continue milk consumption into adulthood and regularly digest the milk of other mammals (most of the other standouts are companion animals).[43] This adaptation provides a nutritionally rich source of calories, proteins, sugars, and vitamins but also exploits the capacity of certain nonhuman mammals to extract nutritional value from foodstuffs that are otherwise indigestible or nutritionally worthless to humans. Although many nonhuman milk producers currently eat grains, soybeans, or maize also nutritionally viable for humans, this practice has more to do with regulatory incentives, market forces, and systemic overproduction of certain agricultural goods in industrial farming than with the underlying capacity of heifers to convert grasses into milk. In the context of humanitarian work, the ability of nonhuman animals to eat grasses and other feed sources to facilitate milk production is far more crucial, and in addition, evidence suggests that in the context of chronic hunger, protein and nutrients derived from nonhuman animals are more beneficial to malnourished humans.[44] Contemporary industrial milk production is primarily a result of the labor of bovines, specifically dairy cattle, who occupy this role because of the efficiencies of grazing on the surpluses created by agricultural subsidies and the volume of their milk production, but also because of the possibility of successfully exerting control over their behavior (which could also be read as agency or political action) at scale and reaping the profits from their eventual slaughter. Many institutions, including factory farming, developed primarily in response to what Deleuze and Guattari would call "lines of flight" or, more concretely, the capacities of nonhumans to exit or resist assemblages of domination.[45] These lines of flight range from sheer physical strength to break away from human handlers to anxiety at the proximity to sites of industrial killing. Milk production would never materialize at a global scale absent technologies of control ranging from breeding and herd management epistemologies to direct control of sexual reproduction, mechanisms such as barbed wire, and many custom experiments in the management of nonhuman animal resistance.[46] Still, specific nonhuman animals are also more conducive to these modes of control and the demands of production. For instance, pigs, another partly domesticated nonhuman animal, produce milk that is nutritious and viable for human consumption. However, their process of milk release, akin to a letdown, occurs only for brief intervals of up to around thirty seconds at a time, making them resistant to the production timeline and demands of agricultural capitalism and physically challenging for humans to milk during earlier periods of pig–human domestic relations.[47]

In addition to its hypothetical nutritional benefits, many other properties of milk make it amenable as a commodity. Milk provides a protein with a complete amino acid profile, a large reserve of dietary calcium, and is ideal for rapid consumption of calories by growing infant mammals. Milk proteins also enable the creation of many secondary, calorically dense, and aesthetically pleasing products, including cheese, butter, sour cream, and yogurt. As a liquid, milk is transportable, relatively easy to store in dried or frozen forms, but it can be enhanced and used as a base for making more complex foodstuffs.[48] Milk is also remarkable because it is a highly reproducible substance that, for nonhumans, can often be created from multiple ecologies, including grasses, in environments that offer sparse nutritional opportunities to humans. It is the capacity of nonhumans to eat in environments less amenable to culturally dominant human diets (fostered by capital and colonialism) and modes of life, to make useless excesses into sustainable foods and commodities, to serve as hosts, making it possible for humans to engage in niche construction, that makes nonhuman milk (and meat) ideal for humanitarian purposes.

Goats and cows have coexisted with humans as part of what James Scott terms "the domus" for thousands of years, and forms of pastoralism likely reach deeper into history.[49] Using nonhuman milk for foodstuffs occurred at some point in this process, although the exact origins of the practice, like many aspects of domestication, are contested and uncertain. For humans, surplus nonhuman milk offers multiple potential benefits just outlined. However, if bovines and caprines are well suited to milk production, it is also because the domus, social life with humans, has primed specific lineages of cows and goats for milk production by shifting their social, biological, and ecological habits to varying degrees. These practices range from intensifying herd behavior and reducing the stress of encounters with other large megafauna, including humans, to modifications in brain size and the shape of the mouth and teeth.[50] Grazing, herding, and interaction with humans are long-standing adaptive changes, impacting generations of nonhumans, but also represent swift plastic divergence at the scale of species evolution. This process aligns with agrologistical imperatives because it transforms ecological otherness, scattered, rhizomatic grasses inaccessible to human digestion, into a means of extending human life. The movement of goats and cows follows from expansion of this mode of agrologistics through the exportation of the modern capitalist state-form under colonialism, with its emphasis on the production of surplus and the disposability of human and nonhuman life. As with dogs and rats, cows and goats are global because of their coexistence with (and possible subjugation to) human ends as part of what Jairus Grove terms the "great homogenization" that has characterized the state, capitalism, slavery, and colonial projects for most of the past five centuries.[51] While goats and cows initially provided the distinctive capacity to produce milk in environments inhospitable to a sustainable diet for human communities, what now distinguishes these species, in particular, dairy cows, is their capacity to produce milk while enduring multiple forms of violence, stress, forced reproduction, control, and slaughter at the scale of nearly an entire species in a process that Derrida calls their "interminable survival."[52]

Nonetheless, cows, at minimum, negotiate, deliberate, and can opt into specific milking practices.[53] Indeed, the evidence that goats, cows, pigs, and other farm animals have complex social, emotional lives, including numerous forms of dynamic communication and engagement, is quite strong.[54]

Given the dynamic process, propensity for bonding, and intergenerational relations presupposed by milk creation, these capacities should not come as a surprise. Moreover, they point out that understanding dairy production exclusively as a result of unilateral force risks diminishing multiple forms nonhuman agency in producing milk, especially in contexts beyond industrial agriculture. While scientific investigations typically interpret caprine and bovine domestication as an outcome of human predation, the transition from predation to management also involved significant changes, with some arguably beneficial to nonhumans when framed in terms of anthropocentric reason.[55] Moreover, the enhanced sociability that accompanied this process was a precondition of the anthropocentric feeling, the fuzzy sense of connection and affective awe, that accompanies nonhuman–human interactions and that enables the aesthetics common to humanitarian interventions with goats and cows; otherwise, these nonhumans would never acquiesce to life in farming institutions. Though in many ways, making milk is always a cooperative endeavor for mammals, initiated because of the differences implied in sexual reproduction and the resulting relationships between infants, juveniles, mothers, fathers, and surrogates, human uses of nonhuman milk constitute an experiment not just in domestication but in the possibility of extending this relationship of surrogacy, using the biological capacities of nonhuman mammals to human benefit, but arguably also potentially easing or regularizing specific pressures facing bovines and caprines, such as predation by other nonhuman animals and inconsistencies in foraging.[56] Milk production could thus be understood in Donna Haraway's terms as an effort to eat or make a meal together or, according to Michel Serres, as a kind of parasitism where megafauna parasitically consume plants and grasses only to be, in turn, the parasites of human consumers of milk, who are, ultimately, parasites in the capitalist extraction of surplus value or, alternately, in a relation where subsequent generations become parasites of prior ones.[57]

Certainly, in the context of industrial production, the heterogenous features of milk, the interspecies breeding, pregnancy and labor of female cows and goats, involve mass violence and homogenization to create the pale, uniform liquid familiar to grocery stores in the Global North. Indeed,

there is an enormous intersectional, medical, social, and feminist literature on milk in the context of human production and consumption.[58] For
bovines, however, almost all studies of milk production and consumption
occur in laboratory settings in the context of large-scale agriculture.[59] Most
of this research, filtered through the prism of anthropocentric reason,
focuses on how to best coax nonhumans into making the most milk of
the best quality at the lowest cost. Major debates concern questions such
as the quality of feed or the use of hormones and antibiotics. In contrast,
precious little attention is paid to the experiences of nonhuman producers
of milk or to the forms of ecological entanglement this process involves,
save in the form of small-scale, more animal-centered farming. Humanitarianism may change little about the basic process of using nonhuman
milk, but it does reflect major breaks with some of the dominant forms
of the intensity, scale, and treatment of nonhuman animals common to
industrial agriculture, which result not only in violence against nonhuman
animals but in unprecedented damage to the biosphere.[60] In this sense,
humanitarian organizations like HI articulate a shift in the context of
animal treatment and management at odds with some of the mechanisms
of homogenization and control common to global agriculture.

The question of eating and eating well is one of the few problems
likely integral to any living thing that requires energetic inputs to thwart
entropy.[61] It is a transversal problem or question with a multiplicity of
potential answers. Humanitarianism's repurposing of nonhuman capacities
for milk makes a material-semiotic shift that is notable because it frames
the process of making milk in different terms than do existing accounts
of the dairy industry. In doing so, humanitarians doubly articulate bovines and caprines as economic resources, as animal capital, as tools for
overcoming the nutritional deficits in subsistence, but that nonetheless
require a sensitivity to the role that nonhumans play in sustainable, vibrant
ecologies and to matters of nonhuman health and welfare. Milk, in effect,
becomes an object of care and concern, but it also functions as a gift, one
that generates the possibilities of sustenance and economic development,
a gift that paradoxically involves humans, soils, and economic networks
but that frequently does not include the nonhuman animals that generate
the milk that forms the substance of the gift.

PROTEAN PROTEIN

Given its nutritional and economic value, it is no surprise that humanitarianism prizes milk as an agricultural product. The historical presentation of milk has varied along with the consumption and marketing of milk and other dairy products. For decades, in the United States and elsewhere, milk has been defined as an essential part of childhood nutrition, a point of considerable dispute.[62] However, milk is not for everyone. As a matter of popular science, lactose intolerance, a reaction to one of the many composite sugars in milk, makes milk indigestible or unhealthy to a large portion of the planet's human population.[63] Moreover, milk is far from culturally neutral foodstuff. While the ubiquity of milk as part of American and European diets has grown during the past century, much of this shift has to do with addressing crises of overproduction in dairy farming, marketing a specific image of family life, and shifts in nutritional epistemology.[64] In this milieu, milk symbolizes a food of nourishment and care. These associations arguably grow from earlier agrologistic motifs and constitute a kind of gastronomical political theology. For instance, the Bible refers to Canaan as the "Land of Milk and Honey," whereas the Milky Way galaxy derives its name from Greek mythology regarding the gods splattering milk across the cosmos. Because milk is primarily a food of infancy, it also becomes a subject of governance as it is implicated in debates about the healthy development of children. Milk governance implies the existence of parental responsibilities overseen by extended families, communities, schools, states, and international organizations.[65] Milk's social function exceeds its nutritional or economic properties to become a matter of care, cultural politics, and biopolitical governance.[66] However, as Vasile Stănescu demonstrates, the promotion of milk and meat as parts of a healthy diet also plays a role in the production of racial hierarchy as part of discourses on white supremacy. Today, alt-right movements point to the existence of lactose intolerance and the benefits of milk consumption as supposed evidence of the superiority of whiteness.[67] In doing so, they build on prior colonial periods in which dietary distinctions were used to observe and legitimate forms of racial hierarchy. The cultural politics of milk are thus also deeply entangled with racialized disciplinary and exclusionary power. These sociocultural discourses on milk also influence the humanitarian context, where the recipients of aid are disproportionately from an implicitly racialized Global South, described in ways that share many of the features of childhood, such as innocence or powerlessness, and are recommended to use milk and dairy production as a means of transforming their economic status and improving civil society.[68] Milk's value is multiple: it secures symbolic associations of cultural identity and colonial power, materially supports a protein-rich and high-calorie diet, and functions as a commodity for economic development.

While humanitarian organizations shift many of the aspects of the assemblage of milk production, they also reproduce the content of the cultural motifs surrounding milk as a food of economic well-being and nourishment as well as the historical practice of recommending dairy production, based on a nutritional rationale, as a technology of the self. However, they frame dairy farming as a donation to the Global South. Given the health hazards of milk consumption, there is a danger of a kind of digestive discrimination in this practice, but one that has been increasingly addressed over time as the problem has become more clearly understood. More importantly, the humanitarian uses of milk redouble its importance as a sign of care and concern. This intensification defines milk's value in relation to both the financial functioning of agricultural markets, a collective good to be secured and managed, and political governance. Because nonhuman milk facilitates both ends, linking nonhuman animals to developing communities directly promotes these objectives. This model of life is not politically neutral. To use the language of Deleuze and Guattari, it links not only flows of nutrients and milk but cultural habits, patterns of language, and modes of existence by constituting a particular type of self, a nub of labor, eating, and transacting processes, as integral to personal and economic well-being. Milk becomes not just a commodity but a means of what Foucault calls "governmentality" or the "conduct of conduct" as it establishes a site for normalizing a specific model of life in the Global South.[69] As Foucault argues, the intention of this practice is somewhat beside the point if the practice, both discursive and material, produces shifts in the self-governance and activity of the

population. In this sense, the practice affects not just hunger but simultaneously, with and without intention, a form of life that it marks as an
otherness in need of transformation. The discourse of the gift signals a
process bound up with what Deleuze and Guattari call an "incorporeal
transformation."[70] Gifting operates as both freedom from a certain form
of want and wariness but also a call to change one's conduct in relation to
others. Animal stewardship, as part of this process, operates as a means
of steering people into a different form of life. The pastoral power of this
gesture, while arguably self-evident, occurs because the goal of gifting
is not only to provide foodstuffs but to incite specific patterns of desire,
models of ethical conduct, and better economic consumption as ideal
outcomes for the beneficiaries of their aid. In other words, milk not only
nourishes the biological body but also becomes part of articulating a vision
of an enriched, political life.

HUMANITARIAN FARM AND
THE CONTINGENCY OF SUBSISTENCE

The stated goal of teaching people to farm is to address chronic hunger,
poverty, and disempowerment, which become acute in crisis conditions.
The fact that this constitutes a form of governance does not mean that
it is not also an admirable aspiration that directly improves many lives.
According to recent estimates, more than a hundred million humans experience serious hunger each year, with tens of millions more subject to
inconsistent conditions on the verge of crisis.[71] Supply chain disruptions
during the Covid-19 pandemic have only worsened the problem.[72] Hunger
and malnutrition create many well-documented short- and long-term
health problems and generate cycles of permanent inequity and debility.
Human hunger is a result of political inequity.[73] Current agricultural resources likely create sufficient caloric and nutritional surpluses to support
a human planetary population of more than ten billion.[74] It is the distribution of calories, knowledge, and economic power, not ecological carrying
capacity in a global context, that likely produces malnourishment. Local,
regional, and multinational challenges, including armed conflict, climate
change, and the disruption of food prices, account for a large portion of

the acute hunger, but these social arrangements are themselves a result of much broader assemblages of capitalist distribution.[75]

By providing nonhuman animals, humanitarian organizations like HI address the root of the problem by enhancing agricultural production at a local level. Given the unlikelihood of major structural, economic, or geopolitical change, it participates in what Sam Moyn terms a kind of "minimalist utopia," one that does not advocate revolutionary shifts in the underlying power structures and inequities but engages in a kind of harm reduction, centered on empowering individuals and communities, to deal with the nefarious consequences of these dynamics.[76] Dairy production does not immediately redress armed conflict or combat climate change; rather, it generates resources to lessen the impact of fluctuations of food prices and local surpluses that buffer communities against swerves of global warming. Moreover, by creating sustainable profits from farming, it offsets the impact of conflict and other disruptions to economic development. In this sense, the purpose is less an effort to reduce hunger, which is a symptom of a broken system, and more an attempt to slowly alter the underlying social-economic conditions that sustain hunger in each community. In doing so, this approach isolates a specific model of ecological relations as the key to altering these conditions. Put differently, the danger this style of humanitarianism addresses is not just underdevelopment or inadequate access to food but a mode of existence called subsistence.

The domestication of animals, farming, economics, and community are tools to extend agrologistics as a means of redressing this problem. The turn to dairy farming (and other modes of agriculture) occurs as part of a broader fantasy of land, space, and stability associated with the stability of contemporary agricultural practice. It is a kind of dream, present even in Dan West's initial visions, of bringing the American midwestern dairy farm to a rebuilding Europe. Drawing from the context of the American Midwest to articulate a model for political and economic relations is a long-standing practice. Midwestern home making, homesteading, farming, and settler colonialism have reworked gender, race, polity, sex, labor, education, military practice, and economic institutions in multiple contexts.[77] As Renisa Mawani argues, "what the politics of cattle tells us is that imperial struggles over land, labor, and resources were never human alone. They were deeply intertwined with the natural world, in ways that positioned animals as willful and disruptive agents of empire."[78] The dairies and farms that sit in the midwestern plains could themselves be understood as part of a durable colonial fantasy, one formed through the technological invention of barbed wire, Indigenous genocide, the cultivation of monoculture agriculture, rigid gender relations, human drudgery, and arduous nonhuman labor. In parallel, this context also produced a signature form of civility that emphasizes a disposition toward charity and pleasantness as consistent features of an implicitly racialized version of social life.[79] These elements also deeply resonate with the project of giving cattle, as if cattle, farm fields, and the Protestant work ethic are the ingredients of a better version of life. Bovines are an integral piece of this fantasy landscape, much as they are also the bodies that materially create the milk and other commodities that supply economic livelihoods.

This model of humanitarianism inherits much from this fantasy. It presupposes that land is currently unoptimized, that this is a failure of economic development (although the fault lies more often with historical relations of global power than with individuals), and communal politics, a representation that dates at least as far back as John Locke.[80] Here subsistence constitutes the outside of civilized life. Although more technically precise definitions of subsistence exist now, the problem of subsistence emerges as part of a hierarchy that grades civilization by level of agricultural production in relation to a Eurocentric, teleological baseline where producing a surplus becomes a sign of an elevated status.[81] This model is increasingly out of step with contemporary historical and archaeological understandings of the diverse ways humans and other species cultivate calories and dispense with surpluses.[82] Subsistence is deeply relational, defined more by the coordinates of capitalist production and technological teleology than by an existence of nutritional or caloric deprivation. Discourses on subsistence articulate a fear of nomadic and migratory forms of life and obscure how the stasis facing large segments of the global human population results from contemporary, violent changes in fusing relations between people and place.[83] Put differently, subsistence is the other of the proper relations of life, labor, and land. Although conditions of contemporary subsistence certainly entail persistent precarity and are inequitably distributed across the Global South, precariousness is only a part of the problem posed by subsistence.[84] Eating always involves precarity. Mammalian bodies are topological donuts, and ingestion is a process of drawing otherness into the body, a folding of the outside, including innumerable agencies, to form the inside of the body.[85] The process of ingestion shows that the body is never unitary but bloated with nonhuman others. Production of foodstuffs similarly depends on a host of climatic, ecological, and increasingly toxic residues that complicate consumption.[86] The problem with subsistence is thus not precarity, in the sense of being dependent on something else, but a particular relation to a specific kind of contingency. Ecological otherness, nonagricultural models of consumption, even acceptance that life may not persist indefinitely all become an outside to the agricultural project. In this way, this practice of humanitarianism differs in its response to subsistence from other humanitarian interventions. Nonhuman animal foodstuffs are not designed to temporarily ward off the contingent effects of desertification, war, or another loss of agricultural production but an attempt to address not individual contingencies but the very possibility of contingency by intervening on the form of life. Bovines, caprines, and, originally, the agricultural model serve as part of a fantasy that tightens human relationships to a predictable future, one secured against contingencies.

Indeed, contingency is intolerable from a humanitarian vantage point. One way to understand this problem, following Giorgio Agamben, might be to read discussions of subsistence as an articulation of bare life, devoid of any protection and exposed to potentials for death without a commission of homicide.[87] In this reading, subsistence involves a mode of existence abandoned to the elements, beyond the pale of law, and subject to whatever powers might assail it. As Agamben correctly argues, bare life is often a presupposition of humanitarian operations. In this framework, modes of life reduced to nutritionally deprived bodies, mainly situated in the Global South, constitute the precondition for acts of political power that govern the destiny of the hungry. People are included in a global political order only insofar as they exist as mouths to feed, a model that frequently serves as a pretext for making lives fungible. However, there is still a deeper issue here that the framework of bare life, with its multiple problems, does not address.[88] As growing evidence suggests, most human existence involves multiple, interacting models of caloric consumption, including hunter-gathering, pastoralism, and scavenging, each occurring in a distinct ecological niche with distinct multispecies relationships.[89] Each of these strategies confronts the problem of contingency using a different set of technologies and different modes of nonhuman coexistence. Technology itself might be understood as a prosthesis that seeks to reveal new ways of reducing a particular type of contingency.[90] It is thus not as simple as contingency, bare life, but the *form* of contingency that creates anxiety in humanitarian discourse, as if the prospect of a life not defined by the existence of a surplus is an anathema.

There are deep ironies with this anxiety about contingency. First, at a planetary level, caloric production is not a problem. Despite varying accounts of ecological limits, considered as a global population, human communities currently create enormous surpluses, many of which feed cattle, chickens, pigs, and other nonhuman animals that are slaughtered to make meat for the purposes of dietary enjoyment, but these surpluses, through various mechanisms of distribution, never change the condition of subsistence.[91] Subsistence is a product of relations of desiring-production, but it is understood in humanitarian terms as tethered to a mode of life marked by privation and underdevelopment, a remnant of a bygone age.[92] Subsistence is rarely framed as a meaningful form of disengagement or resistance to the incorporation of capitalist models of ecological relations; rather, it is presented as a derivative form of life that acquiesces to contingency and, in doing so, exists as the opposite of the good life, which is defined by permanence, presence, and the reserves offered by agrologistics. Given that contingency is constitutive of embodiment, if not a metaphysical property of the cosmos, the revulsion of subsistence is a horror in response to a kind of "blasphemous life," an anxiety about a kind of regression into the unknown in which one would establish a persistent relation to contingency; otherwise, cruelty and suffering would be linked to the underlying social structures that perpetuate them.[93] Condemnation of subsistence is a contemporary reaction to a condition fostered by inequitable political relationships, which fails to consider not only the contributions of modernity to this predicament but the possibility that subsistence is not a form of deprivation but an opposition to dominant modes of capitalist and colonial power and, moreover, that it entails a different ethical relationship to the inevitability of contingency as a problem confronting both human and nonhuman life.

The second irony is that, by turning to nonhuman animals as a means of escaping subsistence, humanitarianism does not seek to eliminate contingency per se so much as affirm that people have a place in the human community, reassuring those who live at a subsistence level that they are not unwanted by producing a relationship of control over nonhuman life. Managing nonhuman animals becomes, in effect, a method of signifying one's participation in a more equitable humanity. However, agricultural work is quite frequently inequitable drudgery, yet even this drudgery is valuable, because despite the discontent it creates, it mollifies the anxieties produced by contingency.[94] As Nietzsche argued in his critique of the ascetic ideal, a will to embrace painful consistency based on the promise that meaning might emerge from pain is historically desirable if it wards off the sense of contingency, void, or meaninglessness associated with existence.[95] Mark Duffield has similarly demonstrated that humanitarian efforts hinge on producing resilient, self-reliant subjects capable of grappling with the structural crises endemic to global capitalism. In this regard, humanitarianism becomes a disciplinary effort to produce a particular kind of economic subject rather than contest the conditions fostered by systems of global inequity.

Here the final twist is that it is cows and goats—not human ingenuity, but nonhuman cohabitation and the use of nonhuman labor—that ward off contingency. Put differently, it is not anything special humans do to elevate their lives away from the otherness of contingency that defines subsistence but nonhuman animals that help to rescue, as it were, humans from the possibility of this condition. Scripting the response to subsistence as an intervention of nonhuman-based agriculture treats cows and goats as that which determine the possibility of humans determining a value to their lives. Put differently, to move beyond "blasphemous life," one must become an agriculturalist, a subject of pastoral power, a shepherd of animals. The nonhuman animal becomes a complement to a proper mode of life, one that both must be managed as the proof of one's escape from subsistence, erased from the gifting and exchanges that facilitate the exit from this condition, but also capacitates, through its own labor, humans' ability to make this transition. Milking another animal provides an ontological reassurance of a human's position in relation to subsistence, where subsistence marks not a lower phase of agricultural production but a model of life poorly suited to the contingencies of life intensified by contemporary capitalism, which creates surplus, uncertainty, and inequity. That subsistence involves myriad complex modes of eating, that deprivation is a product of the very conditions that are treated as the antidote to subsistence, indicates how difficult it is to differentiate the baseline damage to ecology, ethics, and multispecies relations in an age where capital, the state-form, and global hierarchy are ubiquitous.[96] The erasure of the nonhuman resonates with an effort to establish the human in a position of stability, a stability that is not material but ontological, grounded in the hierarchy of managing animals as proof of one's elevated humanity. By providing milk, cows and goats not only rescue humans from conditions of malnutrition but also open the possibility of transcending the contingency of merely subsisting, of merely living. It is the dependence on nonhumans for these tasks that explains why they simultaneously are the source of the gift of milk and erased as participants in this very act.

CANNIBAL HUMANITARIANISM

Perhaps more than any other humanitarian organization, HI, which initiated this humanitarian practice, organizes its work around the theme of the gift. "Passing the gift" is HI's first institutional "cornerstone." A commitment to gifting is emancipatory. It launches networks and generates economic justice, feeds families and creates livelihoods. While the gift takes the material form of nonhuman animals, knowledge, infrastructure, funds, and other adaptive tools, it also takes a social form by freeing people from a condition of subsistence. Gilles Deleuze argues that the way a problem or concept is articulated invites its own solutions.[97] Here the problem of subsistence invites solutions related to the deficits in resources, knowledge, and agriculture. But if subsistence is a symptom of a different problem, such as structural inequities in global racial hierarchies, capitalism, nested regimes of authoritarian power, the postcolonial condition, and so on, then the solution of giving cattle or goats does less to remedy the situation. In contrast, if the fear of subsistence is existential, about an inability to imagine life otherwise, then simply providing resources will never be sufficient. HI's efforts have no doubt improved the opportunities of millions of people, but the gift also involves moral, political, and economic implications that accompany the multispecies assemblages that collectively take risks to address the inequities of subsistence.

All of this, farming, aid, gifting, the escape from subsistence, builds from the underlying capacities of goats and cows. These nonhumans are paradoxically especially prominent in HI's materials, placed in the foreground because of their propensity to generate anthropocentric feeling, and yet only partially present in the discursive economy on the practice of gifting. Cows and goats are a parergon, a supplement, an absent presence in the formation of the gift.[98] Without their life and their labor, there is no milk, no fertilizer, fewer crops, no farm, no network effects, and less opportunity for community and development. It is a partition of the humanitarian sensible, to use Rancière's phrase, that allows nonhumans to make milk and affords them sufficient agency to generate positive humanitarian feelings but limits their role as progenitors of gifts. If these nonhumans did not make digestible milk from the indigestible, then the "gift" would be a burden. Nonhuman labor is the condition of possibility for this form of gifting.

While HI and other organizations do not rationalize or explain their discourse of the gift in relation to nonhumans, there are many possible reasons why sentient mammals might be excluded from the economy of the gift: they lack awareness, meaning, language, consciousness, and so on. In each case, the gift is a human symbolic addition to already occurring material and economic practices. This symbolic addition obligates humans to new ethical courses of actions. So, it is the human *purposing* of nonhuman products that constitutes a gift rather than nonhuman labor. As with all gifts, it is the intention, "the thought that counts." Perhaps it is the theologically inspired faith that humans were meant to shepherd nonhuman animals, a belief in the bones (or fields) of agrologistics, or the desire to emphasize that it is humans who destine nonhumans for use, or perhaps the habituation to farming, a kind of being domesticated to domestication, that promotes this ethical model. Whatever the rationale, it remains starkly anthropocentric, establishing a distance between goats, cows, and humans, between the gift and its conditions of possibility.

As the previous chapter argued, following Derrida, the absence of an awareness of the effects of the gift may, paradoxically, be the only condition that ensures that a gift is a gift because it avoids a parched relationship of debt and reciprocity. Here the ethical problem can be pushed even further. The gift of nonhuman animals is not unconditioned but explicitly linked to an act of future gifting. Paying it forward remains a form of debt or obligation. This process swells the economy of the gift until gifting originates food, farm, livelihood, economy, and development. The gift is the source point, the interruption, that produces these other benefits. Ultimately, the gift is not just a material good but also a way of demarcating the exit from subsistence. If one can gift, then one has produced a surplus, and this surplus is the barrier against contingency, the enacted distinction between subsistence, the other of agricultural humanitarianism, and something better.

Though the gift is a symbolic articulation, it presupposes other material relations. There is a difference between subsistence and forms of agricultural life. Gifts presuppose a surplus, which can be sacrificed. The gift's dependence on this surplus returns the economy of the gift to the problem of nonhumans. As the second section of this chapter argued, milk consumption is a result of the capacities of goats and cows fostered through coproduction (often violently) with humans but, absent a shift to an entirely plant-based model of consumption, wholly dependent on the body and labor of nonhuman animals. The dependence on nonhuman labor creates a situation where either humans define the gift as exclusively human and preclude nonhuman animals from symbolic interaction or nonhuman animals constitute, at minimum, one of the progenitors of the gift, and gifting becomes a multispecies process. This latter position would also entail different ethical responsibilities toward nonhuman partners.

The first position places HI and similar humanitarian efforts in a position of explicitly insisting on human primacy. Nonhumans, like many

humans, do not always labor willingly. Milking occurs in a zone of tenuous power and resistance: goats bleat, cows kick, and so the process is far
from uncontested. In this sense, the gift may even appear as something
closer to what Derrida calls the "worst," a form of erasure that entails
the destruction of nonhuman life for the benefit of human well-being, in
which nonhuman protests are treated as simply nonexistent.[99] Here Derrida's argument may fit better in the context of industrial agriculture, in
which nonhuman animals are killed as little more than "abstract referent,"
devoid of any meaning or significance other than the generation of meat.
Although HI and other humanitarian efforts instigate a break from this
dominant model of capitalist agriculture in situating nonhuman animals in
broader ecological contexts, milking and slaughter both remain aspects of
this practice, and the "gift" is also partly the body of nonhuman animals.
In this sense, there is still an anthropocentric violence operating in the
midst of this ethical project. This violence creates a paradox because
this model of humanitarianism both makes nonhuman animal bodies
the substance of a gift and positions these same nonhumans as emotive,
companions cooperatively building new possibilities for life. If nonhuman
animals create the possibility of gifting in this form, if their companionship, materiality, and labor are part of what facilitates the building of
community, then they also produce the possibility of humanitarian virtues.
If it is nonhuman labor that creates the gift and places cows and goats on
the cusp of working as active participants in humanitarianism, then, in a
twist of this discourse, humanitarian organizations also come perilously
close to recommending that the recipients of humanitarian aid directly
consume their fellow humanitarian actors. In this version, discourse on
nonhuman animals that acknowledges nonhumans as humanitarians would
paradoxically advise mortal violence against humanitarian laborers using
humanitarian principles.

If goats and cows give, if they form the bedrock of a symbolic world and
self-determined humanitarian identity, then humanitarian organizations
come close to making a recommendation that might best be described as
"cannibal humanitarianism," because they would be offering humanitarian
actors as the subjects to be devoured, fully giving themselves up for complete consumption on behalf of the other. Cannibalism, of course, is the

other of virtually every Western discourse on humanity because it violates the foundational tenet of any universalist model of humanism in which the status of simply being human should safeguard a person from violation. Traditionally, discourses on cannibalism constitute a means for ascribing maleficent intention and base animal behavior to non-Western peoples who are implicitly distanced from a civilization defined by principles of species commonality. In this case, if goats and cows were understood as contributors to the gift, as humanitarian actors, then the humanitarian embrace of animal slaughter would become a recommendation of cannibalism. A strong species boundary is a prerequisite for this form of humanitarian universalism in which the virtues that supposedly define the condition of being human, when enacted by nonhuman animals (and the organization even promotes them as part of it), no longer matter and provide no protection against labor exploitation and death by consumption. The other of humanitarian practice can only have a human mouth, a human face, and only an insistence on the quality of humanness saves humanitarian organizations from this predicament.

In the example of HeroRATs, in contrast, the framework of gifting positions HeroRATs as rescuers, protecting human others from mines in a condition defined by an absence of knowledge. This very nonawareness is part of what makes the rats' labor into a gift. Here the problem is different. Unlike APOPO, organizations like HI explicitly describe gifting as the spirit of what constitutes their practice but restrict or omit nonhumans from the gifting relationship to avoid the nefarious implications of cannibal humanitarianism. However, it is difficult to maintain that the goats and cows making milk or being slaughtered do not have awareness of their predicament. Milking, reproduction, and killing are intimate processes. In the growing literature on nonhuman death in the context of industrial meat production, there is extraordinary evidence not only that nonhuman animals comprehend the pain and death they will and do experience but that they actively resist it through any number of means.[100] While this study does not involve any direct observation of the humanitarian context, it is hard to argue that the observations of these contestations do not apply across circumstances. In this sense, the situation of humanitarian cows and goats is the inverse of that of HeroRATs: whereas the rats are arguably unaware of the dangers explosives pose to human communities but are celebrated as givers because their labor facilitates a gift of freedom from violence, cows and goats are likely all too aware of milking and slaughter, omitted from the economy of the gift, and left in a paradoxical position of companion animal and protein source.

In this case, the anthropocentric and sentimental aspects of humanitarianism blend together to demonstrate how the existence and contributions of nonhuman animals matter, producing important shifts in the conditions of life and treatment of nonhuman animals, but persisting in practices of exploitation and slaughter. Moreover, this articulation of the anthropocentric problem is not merely a matter of ethical principle among human communities. The protests, bleats or stamping, of goats and cows constitute a communication of protest to the application of anthropocentric power. The enactment of violence takes place not just as a use of force but as a process of communication, silencing nonhuman objections to anthropocentric tendencies. Humanitarian farming, if this practice can be understood in these broad terms, remains in a deeply ambiguous relationship with nonhuman life. When nonhuman companionability suits, it receives praise as a charming companion animal, and nonhuman animals receive more open, caring forms of treatment, but when milk matters, animals become laboring creatures subjected to anthropocentric reason, and when meat is required, nonhuman death enters the field. The ambiguity marking the affordances and virtues of nonhuman animals in each of these three positions produces the tensions that mark humanitarian discourse on goats and cows. The humanitarian farm is not only a bucolic pasture that aids the hungry or buffers disempowered communities from economic or climatic shocks but also a space where ethical relations to nonhuman otherness are communicated and produced in various forms of exclusion, exploitation, and violence. In foregrounding these relations, ethical possibilities still open up, because, after all, many humanitarian interventions simply bring meat and milk produced by factory farming to feed the hungry, embracing the abstract referent of industrialized mass killing to address hunger. Thus, while this form of humanitarianism remains governed by an anthropocentric order that does not celebrate nonhuman animals—as is the case with rats or dogs—where otherness exists primarily in a mockup of the human figure, and in which rule over nonhuman species may be as much an autocratic, even cannibalistic affair of life and death as it is companionable labor, it nonetheless also creates a set of new relations, questions, and possibilities with respect not only to disempowerment and privation but to the treatment of nonhuman others as part of global political life.

4

Humanitarian Politics on a Multispecies Planet

Bobby, Dia, Magawa, and the baby goat—these are singular nonhuman animals participating in humanitarian interventions. It is a widely distributed group engaging in diverse forms of labor and making the seemingly absurd a reality. Dogs that detect explosives, rats that track infectious disease, goats that feed the world and nurture (human) kids—together, they are like a company of nonhuman animal heroes from children's fiction, a genre that leans into the promise of a happy world without hierarchy, without losers, where care springs from encounters with difference, predator and prey eventually play with one another, the leashed speak their minds freely, and the disturbing ends up being a result of prejudiced habits rather than ontological scars.[1] It is easy to attack this image based on its unreality. Dogs, cows, goats, and rats do not gather to ride bikes let alone act in the pursuit of global justice.[2] There is no grand consortium, no formal forum or parliament of humans and animals.[3] Against an image of nonhuman cooperation, reality allegedly pushes the hard facts of behaviorism and intelligence. Humans breed, direct, and train nonhuman animals to perform specific acts, to observe the strict hierarchies of who sits and stays. They dictate who gets treats for pleasure and who gets treats that poison, who goes outdoors of their own accord and who is put down for displaying too much wildness. Reality alleges that nonhuman animals do not speak, that there are constraints to what nonhumans can do or know. Reality accuses the unreal of being an emotional cushion or buffer against the obdurate character of the real. Silliness or joyfulness, especially born of animate camaraderie, spells doom when it confronts reality because such pleasantries are "a veil that must be torn aside in order to reveal the mute and invisible violence that is devouring them in darkness."[4] Nonhuman animals purportedly labor in humanitarianism because of equally naive human intentions to do good. If humanitarians occasionally celebrate nonhuman laborers, they do so as an afterthought or as a sign of their internal propensity for graciousness despite the limitations or stupidity of nonhumans. Reality exhibits a kind of fanged, fatal pragmatism. Those who remain committed, despite the jarring, uncompromising character of reality, to fanciful ideals, like stories about cooperative nonhumans, however empirical, are, at best, foolish and, at worst, pantomiming human psychosocial dynamics so they do not have to directly address them. Reality is uncompromising, if not always cruel—that is how you know it is real.

Much could be said about the expectations that define what counts as politics, ethics, and the real, but this genre of response to Bobby, Dia, Magawa, and the nameless goat partakes in an especially cynical form of anthropocentrism. It dismisses the possibility that nonhuman acts constitute forms of communication, care, or ethics and sneers at the idea of incipient multispecies politics. It is worth considering this line of attack because it offers a powerful rejoinder to claims about humanitarianism with nonhumans. Stripping away the anger and resentment that occasionally foment at the mere mention of nonhuman capacities, the more modest version of the argument poses a question worth considering: isn't this all a kind of overstatement? Doesn't the argument play up the capacities of nonhumans and, in doing so, transform an unexpected empirical occurrence into a wild politics verging on the miraculous?

This is an important question. The real is often used as a standard to judge the animal. It is Levinas's interpretation of Bobby's "real capacities" that allows him to dismiss Bobby's conduct as preprogrammed instinct rather than sophisticated ethical conduct derived from an exercise of freedom. Although it is easy to show the baselessness of these types of accusations, to insulate nonhumans from anthropocentric invective, typically by showing that human capacities are not all they are said to be, it is more difficult to make the case that nonhumans contribute more than generic laboring capacities in humanitarianism without succumbing to a dangerous anthropomorphism and diminishing the degree to which humans coerced nonhumans into this labor. The charming character of companionable nonhuman animals may do as much damage to nonhumans by capturing them in symbolic fictions, dismissing their capacities altogether. To frame the question differently, how do nonhuman animals influence humanitarianism beyond simply adding specialized labor? What difference are nonhuman humanitarians making?

This chapter offers a speculative response to this question by developing an account of how nonhumans influence not just specific humanitarian practices but humanitarianism as a bundle of normative, ethical, and political concepts. It argues that nonhuman labor not only removes explosives or makes milk but also communicates with and contests humanitarian interventions. By describing nonhuman animal contributions to humanitarianism in reciprocal, if not symmetrical or equitable, terms, terms that afford nonhumans the capacity to both affect and be affected by humanitarianism, a model of political engagement emerges that avoids the problems of anthropocentrically undermining or anthropomorphically elevating nonhuman capacities. Fortunately, a growing body of literature on communication with, in and between, human and nonhuman worlds exists in the study of ethology.[5] Drawing on this literature, the chapter makes the case for reading nonhuman and human conduct in humanitarianism as a form of metacommunication in which nonhumans do, through a variety of means, contribute to the formation of humanitarian concepts. A major part of this account, however, is that, like so many humans, nonhumans also frequently express indifference or, more specifically, nonchalance in response to humanitarian labor. Any description of humanitarian metacommunication requires an interpretation of nonchalance as constituting perhaps the majority of nonhuman (and human) reactions to humanitarianism and considers the subversive potential of this attitude. Thinking about the importance of nonchalance may provide new insight into the limits of humanitarian appeals. Finally, reading humanitarianism through the prism of nonhuman metacommunication reframes their significance, exposing the transcendent aspirations of humanitarianism against a model of immanent situations in which different possibilities for flourishing and care subsist. The book ends by taking steps toward an outline of the paradigmatic challenges this creates for humanitarianism moving forward.

HUMANITARIANISM AND COMMUNICATION

Throughout this text, the power relationship between humans and nonhumans has often been presented as uniformly inequitable. While nonhumans provide distinctive ecological, embodied, extended, and affective capacities that create new methods for demining, food aid, medical diagnosis, and affect-imbued advertisement, humans design and initiate humanitarian interventions and adapt nonhuman capacities to these pursuits. This description treats humans as exclusively determining the meaning of humanitarian discourse and assigning nonhuman animals role performances within it. Using this framework, the case for nonhuman humanitarianism hinges on demonstrating that nonhuman animals display a variety of humanitarian virtues and, by observing their enactment, arguing that nonhumans should be counted as valuable humanitarian laborers. Here the benchmarks of humanitarianism are explicitly set by humans and defined in relation to a specific image of the human. Nonhuman contributions are judged by how successfully they reproduce this image. One of the main critiques of humanitarianism is that while it presents itself as reducing suffering, it also commits a subtle violence of its own as it sculpts the human and humanitarian as a regulative ideal in contradistinction to base animality. Historically, this ideal also polices ontological boundaries surrounding fellowship, family, community, race, sexuality, gender, class, and ability within human communities. In regard to nonhuman animals, it explicitly orients the politics of humanitarianism toward human ends and away from what Carol J. Adams terms the "absent referent" of mass killing in the industrial production of meat or the "virtually interminable survival" that Jacques Derrida views as capturing nonhuman life in a process that exceeds genocide.[6] Identifying the anthropocentric telos within humanitarianism illuminates, at minimum, a tacit complicity, what Derrida might term a "killing silence," with violence against multiple nonhuman others. By adopting a framework based on human-determined, anthropocentric standards of assessment, identifying a unilateral power relationship between humans and nonhuman animal others, and placing these against a background of massive structural inequity, the argument risks overstating the capacity of humans to define nonhuman lives and undervaluing nonhuman challenges to humanitarian politics.

The previous chapters' observations about the roles that dogs, rats, goats, and cows play in humanitarianism certainly contest the image of the human as the sole source of humanitarian virtue by demonstrating that nonhumans add unique capacities for aid, intervention, and care. Put differently, it reveals that humanitarian capacities are creaturely and assembled, rather than exclusively and inherently human. Nonetheless, an approach that proves nonhumans have humanitarian capacity when humanitarianism is defined by anthropocentric standards and objectives is only so helpful because it reifies an understanding of human agency as empowered to set the final terms for any meaningful discussion of care, ethics, and politics. Despite its pretensions, it defaults to representationalist thinking since it traces nonhuman action using a human figure. As such, the strategy of revealing that humanitarian virtues are widely distributed, just like the strategy of documenting that nonhuman animals suffer or experience pain like humans, relies on the notion that the human and the nonhuman exist in a relation of mimicry that can motivate change. Ultimately, it is an appeal based on a politics of emotive analogy.

This approach has many problems. In the past several centuries, many political efforts have argued that humans and animals experience suffering only to legitimize injuring and killing nonhuman animals and those deemed less than or quasi-human. The question of what constitutes the content of shared emotion and experience is contested and, consequently, grows and shrinks in relation to dominant epistemologies of feeling. Emotive analogy is thus unreliable, at best, and capable of legitimizing violence against those who do not feel in the appropriate way, at worst. Solidarity forged of presumed or imagined emotional kinship is as likely to solidify hierarchies as it is to dismantle them, and visions of emotional proximity to a suffering other often produce frustration, resentment, and violence when those you help remain stubbornly other.

Moreover, while analogies can be powerful, this rhetoric bases its model of nonhuman agency on human agency. It assumes that some type of space exists for public debate, a house of commons where earthlings and spokespersons vie to be recognized as agents of speech.[7] In this model, nonhuman animals remain exclusively a point of reference in relation to questions of humanitarian virtue, humane action, and human suffering. It is I, this privileged human animal, making the case to other human animals to interpret other nonhuman animals as also possessing certain capacities and, using the terms of humanitarianism, articulating a plea to consider these capacities with renewed generosity by extending concern to the manifold dangers facing nonhumans. Even in this case, nonhuman animals do not contest the importance of humanitarian ideals; specific humans (who also must make constant claims to their humanity) do so as benefactors of exploited nonhuman animals. The power relationship that frequently defines humanitarianism, where the privileged speak and act on behalf of community, understood as passive victims of suffering, returns across species difference. Here the problem is less the anthropomorphism, since humans inevitability anthropomorphize just as birds avianmorphize, but the form anthropomorphism takes when it views politics as exclusively made up of human statements.[8]

In addition, this conclusion makes a mistake regarding the origins of solidarity. It presumes that discovering the other's similarity to the self, producing resemblances across difference, incites a desire to act on behalf of the other. Psychoanalysis has long taught that familiarity, intimacy, and proximity with the other is far from unproblematic. Seeing the other as oneself may distort and warp this relation as much as it prompts caring, just as the begrudging sense of existing alienated from oneself may foster existential resentment or hatred of others as much as care.[9] This is not to say that the self and identity are not at stake in solidarity with nonhuman others. As Derrida argues, "there is not narcissism and non-narcissism; there are narcissisms that are more or less comprehensive, generous, open and extended."[10] Instead, what animates solidarity has more to do with pleasure in connection to otherness, forms kinship emerging through engagements with strangeness.[11] Resemblances may be a staging point for this shift, but they are not its sole source. Solidarity does not emerge because the other appears like the self but because, as Deleuze and Guattari put it, "we always make love with worlds."[12]

Describing nonhuman animals as producing humanitarian virtues or as part of distributed assemblages that facilitate humanitarian interventions is important but not sufficient as a strategy for evaluating the contributions of nonhuman humanitarians to humanitarianism. It is a position that maintains the split between humans, who access the signifier, discourse, transcendence, politics, and so on, and nonhumans, because it views the latter as meaningful participants in a project of political justice only because, underneath their hair and hooves, they are like a presumed us. It is an argument that, in seeking to move beyond humanitarian rubrics, paradoxically extends the politics of humanitarianism by appealing to the similarity of the other to motivate action in response to suffering. Instead of continuing with this strategy, a better approach might be to interpret human–nonhuman interactions in humanitarianism as a form of solidarity not defined by commonality or identity across difference but emerging from the potentials in multispecies encounters and communication. This model is not mutually exclusive with the existence of a deep structure of anthropocentrism that precludes nonhuman animals from significant political agency and legitimates mass violence against nonhuman entities. Rather, it affords nonhuman animals the potential to resist these conditions by challenging practices, terms, and concepts that guide anthropocentric articulations of humanitarianism and political justice.

Indeed, what if the failure of communication across species difference is not a result of an alleged lack of communication or an incapacity for symbolic reasoning but instead emerges from instabilities inherent to metacommunication? A rich body of evidence already suggests that nonhuman animals engage in constant communication. Birds tweet, whales echolocate, chameleons camouflage, bees dance, dogs bark, and even forms of life that are not animals, such as bacteria or fungi, chemically signal with one another across ecologies or within other organisms.[13] These modes of communication are not limited to intraspecies negotiation or interspecies threat but abundant with metacommunication. Metacommunication, as defined by Gregory Bateson, is "communication about communication."[14] The classic example of metacommunication is a statement that not only contains direct information about the act being undertaken but also communicates about communication itself. For example, dogs at play bite one another; the bite may communicate direct information that this is a threat or attack, but metacommunication shifts the register from conflict to play.[15] The communicative act, biting, ends up communicating more than its apparent, direct content "this is a bite." For Bateson, communication and metacommunication are not distinguishable based off form or content—both bites that attack and bites that play are real bites using sharp teeth—but they produce functional divergence. A typical communicative act delivers explicit content through structures of articulation (in most human communication, using lexemes, graphemes, etc.) that are intelligible because of periods of historical practice. For example, "I am hungry" communicates a simple but informationally rich demand, whether it is mouthed by a child, barked by a dog, or growled by a cat. Each of these statements also has the potential for metacommunication that may reflexively redefine the communicative process to the point that "I am hungry" suddenly transitions from informational content to a matter of urgency or threat.[16]

For Bateson, metacommunication creates the possibility of double binds in which the content of a message at one level is contradicted by communication occurring at another level.[17] For instance, a parent tells their child to "be creative." In this case, the imperative is for the child to act creatively, but, at another level, the child's creativity results from the parent's instruction and so fails to have the kind of ingenuity the parent is seeking. The child ends up trapped—"am I actually doing the kind of creativity the message demands?"—uncertain how to negotiate the relationship because of the instability fostered by the double bind. Metacommunication illuminates these self-organizing traps and patterns that emerge from within a series of communicative acts. If an act of communication can be pinned down with relative precision, the metacommunicative dimension that determines what an act of communication communicates forms an aperture so that any act of communication can always communicate something other than its explicit content and, in doing so, transform itself.[18] Many aspects of discourse, from irony to nonsense, emerge from the potential of metacommunication and its negotiation. These tensions produce structures internal to metacommunication and help explain paradoxical aspects of humanitarian discourses.

Communication temporarily stabilizes by repetitively occurring in iterative series that generate historically distinctive patterns that vary over time.[19] The apparent consistency of communication is a result of these recurrent loops.[20] However, this structure also means that communication is a volatile, open process because each communicative act retains the potential for recursive variation, where a bit of communication communicates something different from preexisting patterns of communication, reinforcing, transforming, or contradicting their prior significance. There are thus all sorts of factors at play in metacommunication that incite creativity in communication about communication. These factors do not necessarily have to do with the logical content of a proposition but with sound, feel, gesticulation, and other variables.[21] Put differently, communication is ecological, hetero, and autoaffective rather than purely discursive. Moreover, the interactive dimension of any communicative act ensures that there is no position outside of communication that enables a strict determination of the meaning of metacommunication.[22] Put differently, the reason all communication is potentially metacommunicative is that there is, properly speaking, no metacommunication or metalanguage that governs what can be communicated about communication. Nonsense preexists and determines the articulation of sense.[23] Metacommunication forms in the interactive series that constitutes communicative acts and builds from intentional as well as unintentional acts, gestural as well as vocal inscription, noises and signals, all of which require enactment through ecological means.[24] There are no rules or language games governing metacommunication but the formation of temporary structures, characterizing particular systems of communication that distinguish themselves through their self-organization into distinctive patterns but remain incapable of permanently reproducing these distinctions.[25] The anthropocentric machine referenced at the beginning of the book, for instance, constitutes a symptom of the dilemma of metacommunication when it articulates the human–animal distinction.

In addition, metacommunication is not an exclusive property of human communicative acts. Rather, metacommunication reframes the difference between human and nonhuman communication as a matter of degree rather than kind. While most nonhumans do not use a human model of language or speech (although exceptions such as great apes and some companion species certainly exist), human communication is not the source of metacommunication but merely one type of metacommunication.[26] Since metacommunication occurs because of the absence of any overarching rule governing communication, any nonhuman capable of communicating is also capable of metacommunication, and communication is ubiquitous among living things because it emerges from plastic interactions with a broader milieu that are likely one of the few generalizable burdens thrust on life to avoid entropy.[27] Indeed, just because something seems like a "base animal behavior" does not mean it is not communicative in multiple ways. The difference between action and behavior is ultimately itself a means of establishing a distinction between communicative acts using a communicative difference. The fact that a tiger is attacking you does not mean that the tiger is not also communicating to you that it is attacking you and that it is hungry or fearful; this act of communication is either just less important, because your continued existence requires you to avoid its jaws and claws, or is actually occurring in a prelinguistic form that jump-starts your flight reflexes, which are also communicating to the tiger your dissent to the proposed attack. There are no stable criteria for differentiating nonhuman communication from human communication when evaluated with respect to metacommunication.

If nonhumans participate in metacommunication, then greater scrutiny needs to be used in evaluating whether nonhuman communication addresses problems, questions, and issues typically viewed as the exclusive purview of human discourse.[28] Everyday examples show plenty of occasions on which nonhuman communication and metacommunication matter. Dogs bark to indicate play, to satisfy a need for stress relief, and to indicate hunger. Metacommunication, whether spoken or bitten, transpiring in laughter, barking, gnawing, or bleating, occurs in the interactive process that produces the code for decoding what these communicative acts mean in the first place. Communication in diverse forms constitutes a key part of nonhuman resistance to multiple modalities of subjection, control, and killing in global economic processes.[29] Accounting for nonhuman communication and metacommunication thus constitutes a central task of laboring with nonhumans, especially in contexts without the possibility of companionship or with nonhumans that actively seek to avoid involvement with humans. For example, Sue Donaldson and Will Kymlicka offer a brilliant outline of a politics of human and nonhuman animals modeled on citizenship theory.[30] Their account begins with the premise that "animals have variable relationships to political institutions and practices of state sovereignty, territory, colonization, migration, and membership, and determining our positive and relational obligations to animals is in large part a matter of thinking through the nature of these relationships."[31] In particular, they distinguish between domesticated animals and what they describe as "wild" and "liminal" animals, which they view as parallel to independent sovereign entities and denizens of human communities that frequently opt to live outside of anthropogenic spaces. However, understanding the preferences of these nonhumans implies the existence of a metacommunicative process that facilitates understanding how nonhuman animals sometimes communicate their actions. Eva Meijer more directly argues that nonhuman animal communication and metacommunication are critical: "non-human animals express themselves, and these expressions need to be taken into account if we want to adequately address how they have been silenced."[32] There is a danger in simply assuming that metacommunication is the same as "speech," because the gesture of ascribing speech to nonhuman others ends up securing a hierarchy in which it is a limited set of privileged humans who decide what constitutes a meaningful communication, what entities are capable of communication, and how the realm of communication functions.[33] Metacommunication is a condition of possibility for the subsequent formation of rational political discourse or argumentation because metacommunication, as Brian Massumi contends, implies a capacity to articulate or distinguish between multiple meanings or potentialities.[34] Play between dogs, for instance, signifies a potentiality between at least two meanings of the bite: "this is to harm" and "this is to enjoy." Metacommunication thus goes beyond traditional forms of speech and discourse because it is a constitutively open process in which there is no predetermination of either the content or form of communication about communication.[35] In Meijer's account, this model of communication becomes the basis for rethinking democratic engagement "to better include animal agency, and to search for new procedures, institutions, and encounters to further deep interspecies political processes and frameworks."[36] Here the argument is slightly more modest. Metacommunication with nonhumans opens the possibility that the goats, cows, rats, and dogs laboring in humanitarianism do not simply contest humanitarian beliefs by proving to humans, using human criteria within exclusively human discourse, that nonhumans are also capable of humanitarian virtues but also potentially protest the terms of humanitarian capture and humanitarian ideals through communicative and metacommunicative means. Metacommunication traverses the gaps between seemingly elevated human concepts and an allegedly base animal world and opens forms of interspecies engagement that go beyond a traditional anthropomorphic model of politics. There are multiple points in the preceding chapters that highlight how multispecies metacommunication produces crucial changes in humanitarianism.

For example, the first chapter discusses the joy of dogs in the process of mine clearance and contends that this joy is part of what challenges the human framework of the explosive ecology as a place of sad, horrifying contingency. In metacommunicative terms, the dogs are not just happy to be performing a task for treats but communicating a joy in their experience of the environment. The fact that human handlers and communities aided by detection dogs sometimes also note a subtle felt difference in the form of demining labor indicates that the resonances from the dogs communicate at a level that changes the affective disposition of human collaborators in mine clearance work. The dogs are engaged in metacommunication — "this is not mine clearance, this is play"—a shift that occurs as they also communicate the information "there is an explosive here," all transmitted to a human via tail wagging, nose sniffing, barking, and tugging on a lead. This direct communication produces a second level metacommunication in which the burdensome labor of standing on deadly ground becomes a delightful, multilayered sensory engagement. Communication and metacommunication generate new dispositions that allow for a different model of ethical engagement in land mine clearance. By opening this potential, the dogs contest an implicit framework of melancholy that accompanies genres of humanitarian work. In doing so, they challenge an aspect of humanitarian concepts: that the work of justice is linked to the problem of suffering, as opposed to the potential of producing joy though an encounter. This contestation is not at the level of a statement, but it nonetheless negotiates with the force of an implied humanitarian concept that structures how humanitarian work proceeds.

In the case of APOPO, the challenges of accounting for the gifts generated from the rats' olfaction signify a gap or breakdown of human communicative systems. These breakdowns generate a patchwork of acknowledgments, recognitions, and invocations of animal welfare to address this gap in communication. When rats enter training regimes to prepare for land mine detection work or TB identification, they also interact with a human companion. Institutions present this bonding moment as a time of multispecies connection, flush with anthropocentric feeling, and as a vital step toward guaranteeing that rats are effective deminers. However, in creating these paired bonds, in which humans and rats start to associate with one another, the rats are not just becoming accustomed to a given human but also communicating a set of conditions under which humanitarian work may proceed. This is also a means of metacommunication, setting the terms under which certain acts of labor will occur. The fact that a nonhuman animal sets these terms just makes the situating strange. For point of comparison, consider Liisa Malkki's ethnography aid workers in Scandinavia. Malkki shows that humanitarian work emerges from practices of self-formation and an effort to communicate desires about one's need to help children and sets the terms through which predominantly older, retired women create care packages.[37] In Malkki's account, it is the formation of these circles that deviates from the expectations and ideals of humanitarianism. Humanitarianism is not only a siloed worker in a distant field but elderly women enjoying their handicraft in solidarity with imagined children they will never meet. Here simple communicative acts about a desire to aid end up also dramatically reframing the model of engagement in humanitarianism. Like aid workers, the rats metacommunicate about the conditions of humanitarian labor, and out of this largely unspoken exchange comes a new set of possibilities for identifying mines, TB, and the victims of collapsed buildings.

In contrast, the goats and cows sent by HI may seem like generic dairy animals with capacities common to nonhumans living on farms across the planet. Their bleats, grunts, stamping of hooves, and roughshod movements across the farmyard communicate disputes to the agricultural context that defines their labor. However, the baby goat running with the camera is also not comporting itself to the needs of a human gaze but taking pleasure in what it finds on its own terms within the farmyard, forcing communication into an indeterminate place where the shortcomings of humanitarian affective frames become apparent. The communicative element here challenges the assumed simplicity of these farm animals; the baby goat will find its own forms of enjoyment in the process of contributing to humanitarianism. The act "this is milk" thus becomes the source of noise, commotion, disruption, and pleasure that must be hushed to make the communicative series of the "gift" sensible in the absence of nonhuman labor.

These examples are speculative because they involve interpreting communicative acts across species difference. They are, and can only be, provisional, anthropomorphic interpretations, but these constraints are constitutive of all communicative processes. Metacommunication occurs because of the structural incompleteness of any communicative effort, the contrasts of perspective and articulation, whether human or not. With sufficient reflexivity, each speculation highlights how communicative and metacommunicative aspects of nonhuman engagement in humanitarianism protest and condition the actualization of humanitarian ideals. In this sense, the examples are cases of not just nonhuman agency but a particular mode of agency that transforms the model of humanitarianism unfolding in each context. This transformation is what moves specific practices of humanitarianism beyond a static image of human suffering or heroism and toward a more multiple, disrupted, joyful, and variable set of practices. Of course, just as there is no metacommunication, there is no such thing as the animal. So, each occasion of metacommunication also needs to be read not as a statement of a species but as a singular communicative negotiation of humanitarian principles and acts. More important, identifying these as moments of metacommunication provides an avenue for beginning to work out, through interaction, what nonhuman humanitarians communicate in their interaction with human humanitarianism. Attending to these processes offers the possibility for building a non-anthropocentric version of humanitarianism that opens avenues for both human and nonhuman participation. It creates insights into the motive forces that incite and sustain humanitarian work and analytical clues about the prevalence of nonchalance as a response to humanitarian urgencies. Ultimately, framing humanitarianism within a human and nonhuman metacommunicative process may disclose new elements in the architecture of the most basic humanitarian concepts. One of the most curious and important aspects of this framework is exploring how even the apparent indifference or lack of communication by nonhuman animals in response to humanitarian labors nonetheless constitutes a form of metacommunicative engagement.

BECOMING NONCHALANT

Goats, dogs, cows, and rats might communicate joy or protest, but many nonhuman animals laboring in humanitarianism appear indifferent to both their work and, more importantly, humanitarian ideals. In most cases, perhaps the great majority, nonhuman animals do not seem to communicate anything about humanitarianism. How should this apparent indifference be interpreted? One reading is that humans are generally such poor interpreters of nonhuman communication that nonhuman messages are overlooked or, worse, deliberately ignored. Another explanation might be that nonhumans have become accustomed to living lives steeped in so much drudgery that communication about this or that labor seems pointless. A third interpretation is that nonhumans have no comprehension of anything beyond their immediate burdens, that they are "poor in the world" and so have nothing to communicate about humanitarianism. Each of these interpretive moves assumes that silence and indifference constitute a deficit or absence in communication.

However, the apparent indifference of nonhumans to humanitarian labor may not be a result of poor understanding, exhaustion, or lack of comprehension on the part of nonhuman animals but rather a form of communicative nonchalance that emerges from the deep ambiguities that affect nonhuman laborers. Nonchalance is not a failure of apprehension but a resistance at the limits of anthropocentric power. Indeed, presenting nonhumans as indifferent turns this silence into a form of politically reductive passivity. Nonchalance appears like indifference because it involves an ambivalence about the forces affecting a given thing, but this ambivalence entails a more complex consideration of the situation and has more in common with the activity of quiet contemplation than unaware complacence. Read in this fashion, nonhuman nonchalance constitutes a potent means of negotiating the ambiguities of humanitarian communication about nonhuman labor. Furthermore, nonchalance is not a special reaction among nonhumans but a disposition that also frequently occurs when humans confront humanitarian rhetoric. Humanitarian advocates often lament how their pleas are ignored by self-interested, nationalistic, or apathetic constituencies.[38] In response, humanitarianism has increasingly turned to thanatography to motivate audiences and securitized humanitarian interventions.[39] Here nonchalance is understood as an uncomplicated, privileged, self-involved attitude in which a subject assumes a disposition of nonchalance because they can afford to ignore or remain unaffected by humanitarian problems. This is a problematic understanding that interprets nonchalance solely from the perspective of a privileged hierarchy.

In her book on Walt Whitman, *Influx and Efflux,* Jane Bennett describes nonchalance as an art or mode of practice that requires cultivation. This art engages in the intervals of what Bennett terms "influence," or the set of vivacious, affective forces that befall a thing as it encounters a multiplicity of others.[40] According to Bennett, the overwhelming tendency in response to multiplicity is to default to a mode of swift, critical judgment: this force is good, this force is evil. Humanitarianism assumes this model in its unequivocal image of human suffering, which produces ethical dilemmas as soon as it enters into practice.[41] In contrast, nonchalance is an attitude, cultivated and elaborated by Whitman, that facilitates "time for more subtle and complex responses to emerge."[42] Nonchalance is partly a stance, Bennett says, partly an attitude, partly a dwelling in an expanse of space and time in which a form of indifference emerges that reflects the deep ambiguity of one's relations to the forces affecting a subject. Nonchalance supports subtle new capacities, such as the ability to listen and engage these forces prior to rendering moral or apodictic judgment.[43] Bennett uses this disposition to elaborate on the importance of treating influences with sympathy, responding to the influence without immediately reverting to reactivity, and as a key aspect of pluralism, democracy, solidarity, and camaraderie. As she puts it, nonchalance contrasts with thrills of judgment and "is the art of *not-sorting* influences, solidarity is the condition of possibility of nonchalance."[44] Nonchalance resonates with taking note, a practice Bennett, again following Whitman, calls doting, of the infinitesimal, ephemeral, gossamer moments of sensation that subtend the possibility of normative habits of consciousness, morality, and politics.[45] As Bennett highlights, nonchalance involves an embrace of influences, which tug and pull in different directions, prior to any move to sort them into dominant registers of sensation, meaning, or experience. Put differently, nonchalance tarries in the event of a sensory encounter in its alterity. It involves a disposition that refuses the urgency of committing to any specific reaction because it does not decide in advance the significance of an experience. Furthermore, according to Bennett, nonchalance is an important democratic virtue because it entails a capacity to consider without resorting to immediate antagonism.

Nonchalance is not a distinctively human attitude. Instead, in Bennett's terms, it is a precarious filiation between a body and the forces that affect it. Nonhuman animals, which exhibit varying forms of plasticity in their experience, may exhibit nonchalance because they, too, exist in extended, sometimes asymmetrical, sometimes reciprocal relations of force and interaction.[46] While Bennett does not extensively discuss the possibility of nonhuman nonchalance, the interval of apprehension characterizes nonchalance in response to multiple forces and potentials. Following Bennett, the indifference exhibited by nonhuman animals in humanitarianism (and certainly other settings) may constitute a type of nonchalance in response to capture by humanitarian labors. Attributing nonchalance to nonhuman animals in their encounter with humans risks some anthropomorphism, but it does not reproduce the kinds of animism that turn nonhumans into pantomimes of human experience.[47] Instead, nonchalance offers a way of interpreting the emergence of an ambivalence that occurs in response to the predicament of being affected by plural forces when these forces are not immediately comprehensible. In the case of nonhuman humanitarians, nonchalance opens the possibility that the apparent indifference or silence to the questions and problems of humanitarianism is not a simple lack of awareness. Paradoxically, noncha-lance involves an intensification of the sense of things, which produces a peculiar incapacity to immediately respond that, in turn, generates a quiet that gives the impression of predictable, dull patterns of behavior. But brooding inside this quiet is a simmering, potent uncertainty.

By confusing nonchalance for indifference, observations of nonhu-man animals foreclose the possibility that their responses are ambivalent, that they involve openness to potentiality, to metacommunication. The observation relies on an anthropocentric judgment that resonates with biases about nonhuman animal cognitive capacities. However, if Ben-nett is correct about nonchalance, and nonhumans appear indifferent to humanitarianism, then this indifference may constitute a form of noncha-lance that humans, because of their own habits of communication, fail to apprehend as disagreement or disengagement about the conditions, actions, or choices involved in a practice of humanitarian intervention. What nonchalance resists is the imperative to judgment because of the sense of urgency surrounding a situation. Although there are certainly moments when this response is appropriate, the inclination to act produces unintended consequences. In an odd sense, nonchalance might reveal a greater propensity for thoughtfulness about the necessity of interven-tion relative to the ethical calculus that demands immediate reaction. As Didier Fassin argues, humanitarian governance increasingly turns to the emergency as its way of governing forms of life.[48] Here, instead, nonhu-man nonchalance offers a kind of resistance to the reflexes of political thought that promotes this model of exceptional, emergency-oriented humanitarianism. It is, despite its indifference, a mode of refusal that communicates messages about both the specific labors asked of a given nonhuman animal *and* the broader presuppositions folded into this de-mand. Attentiveness to nonhuman indifference is thus already a mode of political protest, one so subsumed by anthropocentrism that it never registers as such, which rejects the temporality and acceleration of gov-ernance implied in humanitarian operations.

Framed in relation to metacommunication, nonchalance operates as an ambivalence in response to distinct communicative imperatives. It is a disposition that communicates a deep ambiguity regarding the merits of communication itself and that forestalls final judgment. In doing so, it creates a position of nonalignment with any of the forces that befall a thing. Consequently, humans and nonhumans that exhibit nonchalance become problems, impediments to action that correspondingly require external, behavioral interventions to prompt them to act out of their apparent inertia because, absent this prodding, their nonchalance inhibits an anticipated or desired reaction. Even if a human or nonhuman carries out a requested action, nonchalance produces a form of metacommunication that rejects that the imperative to the act must be carried out in a certain manner. It commits a form of resistance to power even as it may carry out whatever action the powers that be demand. To be clear, nonhumans may also protest in distress, bleat, kick, or seek to escape a situation and thereby express explicit resistance to conditions of capture or violence in humanitarian or other contexts. But, following Bennett, even in succumbing to a demand or obligation, nonchalance persists in a metacommunicative dispute by refusing to be subordinate to the demand that one must be for or against a force. In this sense, though it lacks the clamor of other protests, nonchalance remains a form of dissent. Ambivalence at the demands of humanitarian labor involves an ambiguity over what the demands and imperatives of this labor are communicating, and furthermore, it remains obstinate on the question of whether these dictates should or should not motivate a response. As a mode of disengagement, nonchalance contests the urgency that structures the model of commonality, sociality, ethics, and action that are the implicit presuppositions of humanitarian concepts. Anthropocentrism reduces nonchalance to ignorance or stupidity to preclude the openness to potentiality that nonchalance involves and the quiet resistance to a demand that it involves.

In human circles, nonchalance is understood as an obstacle to humanitarian action because this attitude hypothetically undermines support of humanitarian causes. Most humanitarian rhetoric intensifies suffering, emotion, and the predicament of the dispossessed not only to communicate a plight but to overwhelm the stubbornness of this deep ambiguity. It is "apathy" rather than constitutive uncertainty, disordered narcissism and rampant nationalism rather than deep ambivalence about how to live in global political systems where powerful forces push in multiple directions, that guide this ironically uncharitable reading of ambivalent reactions to calls for charitable action. In the case of nonhuman animals, nonchalance is afforded even less significance because nonhumans are expected to do as they are told, to follow the order of things, or to fade into an undifferentiated and unimportant background. Certainly there is a danger of treating all expressions of indifference as a nonchalance that resists, but there is also a danger in reducing nonchalance to privileged self-involvement or missing it entirely. By doing so, humanitarianism excludes the possibility that there is a politics to nonchalance to these attitudes, one that calls into question whether the current approach to humanitarian agendas leads to desirable transformations.

METAHUMANITARIANISM

If nonhuman animals communicate divergent responses, including affirmation, protest, and nonchalance, when confronted by humanitarian imperatives, then what does humanitarianism communicate when it engages nonhuman animals? In the case of dogs, rats, goats, and cows, there are three different, if consistent, messages, which, in turn, define three distinct relations articulated between human and nonhuman life. First, in the case of some dogs and rats, humanitarianism articulates a thanks for their gifts. The symbolic debts these gifts create, which cannot be fully communicated, in turn become the basis for recognizing nonhuman animals as quasi-humanitarians. Dogs and rats receive public awards. Here nonhumans are recognized explicitly as humanitarian agents and humane actors effectively extending categories traditionally associated with human conduct to include animal others. Second, in the case of smiling, charming goats and cows, humanitarianism articulates an appreciation for their sociality, endowing them with sufficient agency to be companionable, but leaving them largely aside in broader discourses about humanitarian action. In these examples, anthropocentric feeling defines the relationship to nonhuman humanitarians in which the emotive or affective reactions created by human–nonhuman labor and affability define the role of nonhuman animals in humanitarianism. Finally, in the case of other goats and cows, using the same set of distinctions, human and nonhuman, giver and gift, humanitarianism distinguishes people who contribute to a universal cause, humans who gift and pass the gift, from those entities that cannot gift and so may merely be put to labor and death. In this case, anthropocentric reason transforms the life of nonhuman animals into little more than an "abstract referent" devoid of value, except as a means of perpetuating human life.[49] Put simply, nonhumans are positioned either as companionable humanitarian actor, liminal participant, or nonliving commodity. Between these three positions, humanitarian communication captures nonhumans in a bind. On one hand, humanitarianism communicates that nonhuman life and labor open a horizon of dignity and inclusion as respected members of a self-identified human community on account of the contributions nonhuman animals make to the well-being of this community, and, simultaneously, on the other, nonhuman life and labor are not meaningful save as a persistent resource that requires humans to engage in killing labor. There is thus a deep ambivalence for nonhumans, a rift bubbling from within the indeterminacy of the anthropological machine, legitimating both dignity and death, structuring the possibilities for nonhuman animal participation in humanitarianism.

At this point, humanitarianism does not include any substantive or formal discussion of the relationship between nonhuman agency and humane action. Instead, humanitarianism navigates the twists of the human–animal relationship, attempting, as with previous iterations of the anthropological machine, to provisionally distinguish these terms, only to have them collapse into one another, drawing primarily on dominant discourses about human suffering and animal welfare as well as scientific epistemology. In this sense, humanitarianism recreates a dynamic that, as Cary Wolfe demonstrates, is common in contemporary politics, where nonhuman animals are caught up in opposing biopolitical tendencies of mass violence and dedicated care.[50] Humanitarianism is not unique in preaching compassion for some nonhuman animals while simply admiring or destroying others in the same proposition. Explaining the origins of the historical wobbles and inconsistencies of human–animal difference goes far beyond the scope of this project, but, at minimum, the problem develops because of challenges in articulating even basic metaphysical ambiguities about life or existence.[51] With humanitarianism, which inherits the anthropocentric tendencies of its context, the promises that are made regarding the welfare of nonhuman animals are attempts to pay back a kind of debt for their labor across the communicative gaps that exist between humans and nonhuman others. These attempts continually run into the ambiguities inherent to the articulation of difference with multiple impacts. First, the ambiguous character of humanitarian communication explains the emergence of nonchalance as a disposition of nonhuman animals caught up in humanitarian labor as their existence in a humanitarian apparatus of capture is subject to both predatory power and pastoral care. Second, it suggests that, despite their appeal to universalizing discourse on the human, humanitarian practice with nonhuman animals operates in the form of continuous exception as individual humanitarian agencies employ anthropocentric power to define their relationships with nonhumans through both inclusive violence and inclusive recognition, depending on the singularity of the problems they address as well as the nonhumans they work with. Finally, this ambiguity highlights a few strategies for tugging both broader understandings of humanitarian principles and specific humanitarian practices in more generative directions that support multispecies solidarity and justice.

Nonhuman humanitarianism exposes one of the double binds structuring the articulation of humanitarian ethics. By making human suffering its object of intervention, humanitarian communication indicates that a specific type of experience supersedes other interests, identities, and conflicts: personal, religious, national, social, and ecological. In humanitarian terms, the human, unlike the surrounding "natural world," is special, something paradoxically presumed but not given, a theological remnant, and, hence, a station that requires distinct protections. In a way, humanitarianism is more anthropocentric than speciesist because it identifies not humanity constituted as a biopolitical species but the neglect or mistreatment of individual humans on account of their humanity as its foundational problem.[52] This emphasis depoliticizes as much as it strives to engender caring. Moreover, it masks, in metacommunicative terms, the structural incompleteness of the human as a concept that cannot be specified, that has no *point de capiton* and consequently no a priori sense of the value or meaning of suffering outside of relationships and contestation.[53] This instability drives humanitarianism to constantly supplement the appeal to humanity with other values, practices, and decisions that contravene the shifting ideal at the center of humanitarian thought. The contestability of its concept produces tensions whenever the ideal actualizes and, consequently, generates double binds. These double binds are typically identified when humanitarianism confronts "practical dilemmas," such as resource limitations, the threats facing humanitarian workers, the designs that seek to aid, the urge to tacitly imprison communities that merit help, and the embrace of "lesser evils" no matter how great.[54] If skepticism of humanitarian norms is common, it is partly a result of the fact that this structural incompleteness generates dissonance between the aspirations and actualizations of humanitarianism as a condition of its existence.

The inevitability of the double binds that characterize humanitarian practice is not an alibi for inconsistency or violence. Rather, the double binds indicate a need to analyze the expectations associated with humanitarianism, the mode of humanitarian communication, its model of politics, and the way that structures or systems of value are at work in the articulation of humanitarian concepts. Framed as an iterative, historical, and communicative system, humanitarianism articulates a gesture, common in Western metaphysics, of establishing value by invoking a higher, transcendent meaning. The vertical movement this gesture of transcendence involves structures the topology of historical systems of thought about animality. It is this movement that dignifies the human as political, divines the human in relation to god, separates the *cogito* of thinking man from animal automaton, splits the signifier and symbolic reasoning, grounds the possibility of overcoming, defines second-order intentionality and intelligence, and, even in the most terrestrial, earthly visions, spawns the dream of an escape from spaceship earth. From one perspective, the entire project of the human is a series of different attempts to construct a transcendent exit from the horizon of the animal out of a fear that the animal exists as a kind of untenable, immanent finitude. As Elizabeth Grosz explains, what motivates this line of flight is that, at the same time as the human emerges as vertical possibility, the animal becomes a "reminder of the limits of the human; its historical and ontological contingency; of the precariousness of the human as a state of being, a condition of sovereignty, or an ideal of self-regulation."[55] Becoming human, becoming a dignified humanitarian, is an exit from this animal condition. So many philosophical, political, and ethical concepts emerge from this gesture, this appeal to a transcendent model of the human, that the singularity of a life, defined without the urgency of the vertex, almost entirely disappears.

HUMANITARIAN POLITICS BEYOND THE HUMAN

The ruse of vertical movement likely formalizes when human societies turn to agrological principles, which necessitates a distinction between humans, domesticated nonhuman animals, and the wild, multifarious depths of a planetary (if not cosmic) ecology.[56] But what the vertical movement actually tries to correct is the underlying vertiginous instability that defines attempts to ground metacommunication and separate it from this plane. If the anthropocentric machine seeks to divide human and animal, the instabilities of metacommunication make such a division constitutively impossible but also produce the ruse of transcendence that renews the desire for this pursuit. This ruse is highly persuasive because it gives the impression that it is possible to permanently ground a mode of life that can adequately judge everything spread out on the horizontal plane. Each version of this vertical movement simply leads to a different perspective, a distinct plane of reference that cannot exit its embeddedness in a horizon within an ecology. Humanitarianism operates by folding the possibility of this exit directly into the human condition. The reason this fold creates tension for nonhuman others (and for the humanitarians who want to include them) is that by linking vertical possibility with human life, the gesture makes the question of how to be a good humanitarian inseparable from the question of human–animal difference. Animality, as it were, contaminates the definition of the humanitarian just as humanitarianism, which captures nonhuman capacities, cannot fully articulate or integrate its own excesses. Violence and care, predation and compassion, are the ripple effects of this inward twisting movement of escape. This process has several implications and consequences for humanitarian practice.

First, nonhuman animals are not the only ones who suffer from the double binds of humanitarianism. The classic critique of humanitarian ideals argues that humanitarianism constructs victims as the passive objects of sympathy and power. Consequently, humanitarian initiatives secretly disempower the people they claim to serve. The argument continues that, to establish the distinction between active savior and passive victim, humanitarianism creates a third term, the inhumane, which depoliticizes violence, moralizes conflict, and treats specific humans like evil, inhuman monsters.[57] These three terms or positions vary in form and content, but they constitute tendencies or consistent attractors that emerge from the structure of humanitarian communication. The topology of humanitarian concepts, elevating the human through a vertical movement, generates an instability with respect to the borders of these very terms. When humanitarian discourse starts to speak to the exigencies that define a crisis, when it actualizes, this notion of value creates a constitutive outside, but because of the way the human is folded as both ultimate value and the basis of humanitarianism, this outside also forms in relation to humanity. The human exists as both subject and object of humanitarianism, the thing it acts on behalf of and the source of its danger, both poles of this duality existing in an emerging relationship of reciprocal presupposition.

Humanitarian regimes and interventions are built from a combination of aspirations for a universal protection for human life on account of its humanity and from practical responses to specific problems. This is the lawlike aspiration of humanitarianism, explored in the second chapter, which seeks to articulate a new ground for normative order. However, the topological features of humanitarianism ensure that the protection afforded by humanitarian interventions is always compromised in advance, that the legal aspirations undergo a continuous process of deconstruction, not because they are self-interested or possess insufficient resources, but because their self-articulation creates double binds immanent to the concepts of the human and humanity. Indeed, many critiques of humanitarianism argue that the emptiness of universal aspirations leads to a kind of default toward seemingly more secure foundations for political identity, such as the state or nation.[58] Even if humanitarian ideals displaced these foundations for identity, this shift would not resolve these tendencies because they are borne into humanitarian practice by the topological connections that make the terms operative in the first place. Although humans are safeguarded to an extent because they can, in a way nonhumans cannot, contest their disempowerment by claiming an identity as human or as members of an established political community, the liminal place produced by the structure of humanitarian power creates degrees of exposure and precarity for both human and nonhuman life because both end up captured by structural twists that may contingently license abandonment and violence.

Second, the structure of humanitarian discourse produces a paradigm of inward, even tautological focus. If questions, aspirations, and practices of the good emerge from and return to the human, it becomes increasingly difficult to open connections to nonhuman agencies or to understand the human, however defined, as a provisional result of a network of singular, ecological relations rather than a generic ahistorical condition. Félix Guattari once provocatively articulated a need to think in relation to three ecologies that he argues exist in reciprocal presupposition. Roughly, these ecologies are the self, the social, and the material, or what Guattari termed "a nascent subjectivity . . . a constantly mutating socius . . . an environment in the process of being reinvented."[59] The loop that defines humanitarian concepts occurs in a continuous movement between the first two of these terms and only confronts its entanglement with nonhuman others because of the ambiguity of what constitutes human relation to nonhuman animal difference. Moreover, humanitarianism seeks to arrest the mutations of the socius by articulating the human as a value outside this process of becoming. The third ecology is scarcely ever a source of humanitarian consideration, unless it concerns operational problems or directly impacts human interests. This ecology is, by default, a background condition for human agency rather than a lively series of meaningful actants.[60] Entering nonhuman animals into humanitarianism, poking at their imbrication in ecological relations, disrupts the loops that characterize humanitarian discourse. Expanding on this break hinges on considering nonhuman communication and metacommunication as a significant, and as of now unconsidered, contribution of nonhumans to humanitarianism. The immediate impacts of humanitarianism's anthropocentrism are also profound. In the simplest sense, the focus on human suffering makes humanitarian efforts concentrate on questions of nonhuman life and environmental well-being solely through the prism of how they affect human communities. Climate change, for instance, becomes primarily a problem of human displacement, but the destruction of ecologies, both purposeful and inadvertent, is understood as a cause of this problem rather than a problem in itself. The sixth mass extinction and the purposeful elimination of specific species or ecologies becomes a marginal consideration relative to human need when interpreted through humanitarian prisms.[61]

Furthermore, as numerous activists and scholars have documented, the violence of industrial meat production and other modes of agriculture is startling. To paraphrase novelist J. M. Coetzee, if the current intensity and scale of nonhuman animal destruction are not retrospectively understood as a problem of grave concern, regardless of the question of the ethics of eating meat, then it will signify a depth of thoughtlessness that is, for lack of a better term, unbreachable.[62] Setting aside an abstract, normative debate over whether it is ever ethical to kill human or nonhuman animals, the uniqueness of this historical assemblage, the sheer volume, mode, and intensity of industrial slaughter, with nearly seventy billion nonhuman others commodified and killed for the purpose of consumption each year, with unprecedented effects on planetary ecology, generates problems for even the most cautious, complex, and anthropocentric ethical calculus.[63] Indeed, even if nonhuman others are not considered valuable, judged exclusively in anthropocentric terms, the unintended consequences of this process are staggering.[64] As Derrida points out, the condition is so acute that it leads concepts like genocide and crimes against humanity to falter or become scandalous.[65] One of the most frequent frameworks for addressing this condition is through an appeal to animal sentience and suffering. Historically, in cases such as the treatment of cows and other companion animals, these concerns for animal welfare precede and provide the historical form for the later emergence of humanitarian concerns about figures like children, laborers, slaves, and other passive victims.[66] Yet, humanitarian presuppositions about human suffering distance it from these questions. Part of the reason why the mere presence of nonhuman humanitarians is subversive to humanitarian paradigms is that their inclusion as meaningful humanitarian actors shows that the background of nonhuman destruction, the third ecology, cannot be so easily parsed from a foreground of human suffering, that maintaining the two as separate ethical considerations hinges on a fairly flimsy articulation of human value that obscures the ecological relations within which both human and nonhuman lives subsist.

The third implication follows closely from this second point. The difficulty of articulating the third ecology in humanitarianism, its emphasis on the transcendent value of human life, that is nonetheless historical and immanent, makes it harder to identify what about humanitarian practices encourages flourishing, thriving, and forms of care. The third ecology is a platform for debates not only about whether humans or nonhuman animal needs are worthy of political action but also about the conditions of possibility for action, obscuring the fact that things do not need to become ecological because, as Tim Morton puts it, "they already are ecological."[67] The historical baggage of the human that confines humanitarian discourse errs by searching in the wrong places for care and flourishing because of its point of emphasis. This error also makes it more difficult to engage in humanitarian intervention in the first place, because, just as it limits the field of possibilities with respect to nonhuman animals, it restricts an understanding of agency, its varying distributions, and dependence on ecological relations. Here nonhuman animals communicate in ways that not only demand recognition but make a difference in humanitarian politics. Perhaps humanitarianism may even learn from endeavors in multispecies thriving that do not inherit anthropocentric assumptions, such as the movement toward sanctuaries for farmed animals. Sanctuary entails creating a space that allows nonhumans to escape from conditions defined by commodification, certain death, and continuous destruction *and* an active effort to invent new, if imperfect, conditions for flourishing amid structures of slaughter and ruination.[68]

Even more mainstream studies of human and nonhuman interaction have established multiple types of communication across species difference.[69] Nonhuman communication occurs as part of an immanent, ecological field that humanitarianism struggles to articulate and, because of the ruse of the vertical dimension, is inclined to dismiss. Although it would be a mistake to assume that all communication from nonhuman humanitarians is the same or represents a common sentiment based on a shared essence, it is equally problematic to atomize nonhuman communication, to downplay its potential as politics that involves affiliation and solidarity with unknown others. In both cases, the reason humanitarianism struggles to embrace these forms of communication, even as they embrace practices that require nonhuman labor, is that they involve an alternative mode of political engagement based on a different schema of justice than the one championed by humanitarian orders. Unlike the humanitarian elevation of human suffering to a transcendent ethical principle, a process that presupposes both the value and communicative efficacy of human life, communication across species difference occurs without recourse to a vertex because the process depends on and emerges from its ecological entanglement. There is no common horizon that establishes an anchor point beyond discourse, no metavalue to ground communication. Thinking through this form of communication entails contending with sentient, plural horizons, a multiplicity of gaps of nonsense, and incommensurate perspectives. There are no a priori "universal" or "global" concerns but an interactive process that is fraught, entangled, and coexistent but not fully relational, distended and assembled, that leaves open new potentialities.

Feminist scholars like Chris Cuomo, Lori Gruen, and Donna Haraway have long argued that care emerges from entangled relationships including a multiplicity of human and nonhuman others.[70] The ethical question in any specific entanglement is not just about the quality or forms of labor embedded in care but also, as Maria Puig de la Bellacasa eloquently says, about how to stay "open to an unknown: How many are 'we'? even if this 'we' always involves some aspiration, speculation, and enchantment."[71] Although the need to consider nonhuman communication as formative of humanitarian politics might seem like a gesture toward greater universalism, toward the inclusion of everything and anything, the politics it entails is not so simple. As Thom van Dooreen notes, "while we may all *ultimately* be connected to one another, the specificity and proximity of connection matters—*who we are bound up with and in what ways.* Life and death happens inside these relationships."[72] In this sense, nonhuman communication, first, needs to be understood within its context as a potential contestation of the conditions of labor and life of specific nonhuman others. Yet, communicative acts also cannot be understood as isolated statements. At minimum, they nudge humanitarianism toward reflecting and extending its horizons to include the third ecology, broader questions of multispecies justice, engagement with the Anthropocene as a multispecies problem, and an exploration of new methods of engagement beyond the teleological premise that the future will be defined solely by human possibility and human justice. The ethical possibilities of humanitarian politics hinge on the degree to which specific institutions and actors remain open to probing these limits. Although humanitarian institutions are not capable of unilaterally overturning anthropocentric power, each experimental engagement with nonhumans offers a point of potential solidarity for coalescing around a richer, multispecies mode of collective engagement, where communicative norms and values are not predetermined by anthropocentrism and where interspecies work may result in greater creativity in response to many of the forms of global violence and violation precisely because anthropocentric biases have precluded substantive exploration of these potentials. This kind of humanitarianism, which would move beyond the human, could contribute to multiple, resonant forms of resistance to structures of hierarchy and domination and in which the local question of how to engage with specific nonhuman others links up with collective questions about what forms of equity, ecology, and politics support multispecies flourishing. Moreover, by interrupting the emphasis on humanity, extending toward a constitutive outside, beyond the horizon of human interest, humanitarianism would also have a starting point for rethinking its apolitical premise and admit conflicts in which interconnection involves pluralist conflicts (both within and across species difference) rather than forms of implied unity.

In most accounts of humanitarian work with nonhuman animals, the nonhumans have been brought to labor because of a humanitarian apparatus of capture, but their importance is obviously not reducible to anthropocentric reason, anthropocentric feeling, or a consequentialist determination of the number of human lives saved. Judging humanitarian practices solely from this perspective assumes that human life has global value, as opposed to it being one among an ecology of values.

This assumption is untenable because it is impossible to separate fully from the contexts within which it emerges. Care takes place as both a fragile, entangled process in which value emerges through encounter and coexistence and as a process that influences broader assemblages, packs, and political movements, opening up new possibilities for solidarity, resistance, and revolt.[73]

Here, in the middle of a budding coexistence, nonhuman humanitarians currently aid in a flourishing that dramatically assists their fellow humans. However, what they are doing is not merely humanitarianism but a mutual work of interspecies labor that has yet to fully consider the status, needs, and statements of its nonhuman collaborations. Humanitarian interventions with nonhumans do enact generosity and often lead to the outcomes they are intended to create, but this generosity differs from the type envisioned and championed by humanitarian discourse in terms of the subject caring (human vs. multispecies), the mode (care as empathy vs. entanglement), the process (rational and emotional vs. metacommunicative and ecological), and the result (reduction in human suffering vs. emergent generosity within an assembly defined by an open horizon). These ethics require further work to consider the place of nonhumans, the role of ecology, the place of politics, and agonistic respect for the nonchalance affecting both humans and nonhumans. Indeed, nonhuman animals are not simply miming humanitarian activities or demonstrating their capacity to be humanitarians; they are helping to transform the practice of humanitarianism, causing it to become something different and, in doing so, also changing their human companions in labor. In a sense, humanitarianism with nonhuman animals may become less humanitarian in this exchange but, through this subtraction, find a new set of values and horizons. Where humanitarianism is typically understood as a global, empathic institution defined by commonalities regarding human dignity, experience, suffering, and life, read in relation to nonhumans, this paradigm appears more and more to be addressing only a narrowly defined set of human inequities and interests while also subtly and apolitically structuring discussions about what forms of life ought to be championed and cherished. Instead, humanitarianism with nonhuman others involves a series of partial assemblages of concern, slowly knit together by tenuous efforts at incommensurate communication and forms of joy and hospitality produced by contingent encounters. If non-human humanitarians demonstrate one thing, it is that humanitarianism has always been a poor label for the potentialities of care and generosity that unfold amid a multiplicity of conflictual but coexisting forms of life.

Acknowledgments

This project benefited from the insight, comradery, and critique of many people. The initial notion of exploring nonhuman things in humanitarianism was born out of conversations with Stefanie Fishel. She read the first drafts of what became this book and shared her wisdom on ecology, microbiology, and politics. Chad Shomura and Sara-Maria Sorentino provided intensive and extensive feedback on large portions of the manuscript. They encouraged me to pursue more radical instincts and made me more sensitive to what I had left unthought. My good friend Alex Barder dialogued with me almost daily about biopolitics, racism, politics, and animality and shaped many of my insights into these topics. Siba Grovogui, with his signature humor, shared stories about "Sily" and the humanitarian-like roles played by elephants and other nonhumans. Together, we found great humor in the quirks of Western metaphysics when it came to nonhuman others. Derek Denman first attuned me to the role of dogs in political life and helped me to articulate many of my first thoughts about our relations with companion species. Debbie Lisle was an immediate supporter of this project and regularly connected me to new materials on nonhuman capacities while reminding me to consider the absurdity and playfulness at stake. Jairus Grove added depth to my understanding of neuroscience, sensation, ecology, and the apocalypse across creaturely worlds. Several years ago, Paola Marrati generously read, gave feedback on, and then dialogued for many hours about a far, far too long paper that formed the basis for many of the theoretical maneuvers in this text.

My understanding of violence, law, humanitarianism, and the human–animal distinction was expanded, both directly and more indirectly, through many conversations on the concepts in this manuscript. Many of these exchanges included, in no particular order, Nisha Shah, Helen

Kinsella, Dan Levine, Anna Agathangelou, Dan Monk, Sinja Graf, Tim Hanafin, Bill Connolly, Lauren Wilcox, Jessica Auchter, Adam Culver, Tamara Metz, Nicole Grove, Oumar Ba, Martin Coward, Jishnu Guha-Majumdar, LR Danil, Tim Vasco, Jane Bennett, Justin Eckstein, Cara Daggett, Caroline Alphin, Paul Kirby, David Cram Helwich, Andreja Zevnik, Patrick Thaddeus Jackson, Caroline Holmqvist, Rafi Youatt, Casey McNeill, Jonathan White, Risa Kitagawa, Matthew Leep, Jelena Subotic, Jennifer Culbert, Antoine Bousquet, Kevin McSorley, Hitomi Koyama, Jeff Bachman, Charmaine Chua, Michael Shapiro, Shampa Biswas, Alex Hinton, Brent Steele, Beth Mendenhall, Alison Howell, Louise Wise, Francois Debrix, Amy Niang, Bob Vitalis, Evren Eken, Emily Wills, Mike Baxter-Kauf, Audra Mitchell, Andrew Ross, and many others. Although I never met her, Lisa Smirl's scholarship had a formative influence on my thinking. I also am grateful for the excellent observations and suggestions of the anonymous reviewers of this manuscript, which greatly improved the scope and definition of the project.

There is perhaps no department in the world better intellectually suited to supporting this venture than the School of Interdisciplinary Arts and Sciences at University of Washington–Tacoma. My colleagues in Politics, Philosophy, and Public Affairs, Katie Baird, Jane Compson, Michael Forman, Mary Hanneman, Matt Harvey, Anna Lovász, Bidi-sha Mallik, Darrah McCracken, Amos Nascimento, Emily Thuma, and Charles Williams, all contributed to this endeavor through intellectual solidarity and critical reflection. Although no longer members of our program, I would be remiss if I did not also thank Turan Kayaoglu, Sarah Hampson, and Will McGuire. In our broader school, I owe thanks to Jutta Heller, Joyce Dinglasan-Panlilio, and Jim Gawel. They offered their scientific expertise on matters ranging from genetics and volatile organic compounds to ecological systems thinking. Annie Nguyen helped me to articulate the value of this project to a broader audience. Libi Sundermann gave me excellent advice about selections from this book. I would never have made it through another quarter, let alone completed another book, without the indispensable aid of Jessica Asplund, Maria Hamilton, Shane Agustin, and Hannah Coache. I cannot do justice to all the students who have impacted my thinking about this book through their insights, stories about their interactions with human and nonhuman animals, and questions about this project. My sincere thanks to all of them.

My thanks to Pieter Martin at the University of Minnesota Press, who has been a stalwart supporter and excellent guide in the creation of this manuscript since I first proposed it to him. Peggy Reif Miller graciously provided valuable insight about the history of the Heifer Project. Jen Houser and Cheryl Brumbaugh-Cayford with the Church of the Brethren assisted me in accessing early images of Dan West. I am incredibly grateful to Lily Shalom and Caterina Caneva Saccardo for the generous donation of their time, expertise about APOPO's practices, and assistance acquiring images for this text. Finally, Jihee Shin spent many cold, rainy mornings talking about the sociality of dogs, posthuman ethical quandaries, and interspecies grief as well as providing me with the time to complete this work. She has my gratitude. Many nonhumans also impacted this project and, given the content, it is worth noting several of these encounters, including the dog that attacked me while I was running, particularly defiant exchanges with one especially elusive goat, failed efforts at interspecies communication with two birds that were briefly denizens in our home, and the baby rats I attempted to rescue from severe dehydration. Each of these interactions was, for better or worse, generative of elements of this project.

My family has been a wealth of support and love. Katie Hall and Elizabeth Heatherington have listened patiently to hundreds and hundreds of curious details about nonhuman worlds and helped the rest of us weather the Covid-19 pandemic and the contingencies of life. My mother and father were exceptionally supportive whenever I brought up the book and shared many anecdotes about their and my engagements with nonhuman animals. My sister Mollie should be credited with first pushing me to think about the relation between nonhumans and genocide, a vital stepping-stone in the development of this project. Oliver, Bosley, and Laila, our nonhuman companions, have been participants, inspirations, and interlocutors through the whole endeavor. Indeed, one of the curious things about writing and thinking on nonhuman communication is the degree to which it makes you appreciate the richness and complexity of these daily interactions. Collectively, they helped to refine many of these concepts through appreciation and resistance to my anthropomorphic version of our shared ecology. Emmett has been a constant joy. He is full of discoveries, mirth, quirky humor, and implicit reminders to question any order of things, any presumed partition of the sensible. He also helped format the table of contents. Adira may have talked with me about this project more than anyone else. She is a font of scientific knowledge, an unflinching defender of the idea that fairness should include everyone and everything, and an aspiring wildlife veterinarian. I will forever relish all our time learning and talking together. She aided me in selecting and organizing the images for this book.

But credit should go where it is due. In April 2019, about a month after my previous book came out, Bevin and I went out for a celebratory lunch. The topic of a future book came up, and I told her that I would likely start a project on the role of weapons in genocide but had also thought about returning to the subject of nonhumans in humanitarianism. I made the case for the project on genocide. "That's silly," she briskly replied. "People love to think about animals, you find the subject totally enthralling, and, plus, you will just be a happier person doing that." Simply put, this book would never have existed save for her timely intervention as well as her considerable care, brilliance, and generosity.

Notes

INTRODUCTION

1. Emmanuel Levinas, "The Name of a Dog, or Natural Rights," in *Difficult Freedom: Essays on Judaism,* trans. Sean Hand (Baltimore: The Johns Hopkins University Press, 1990), 153.

2. Emmanuel Levinas, *Entre Nous,* trans. M. B. Smith and B. Harshav (New York: Columbia University Press), 74.

3. Emmanuel Levinas, *Ethics and Infinity,* trans. R. A. Cohen (Pittsburgh, Pa.: Duquesne University Press, 1985).

4. Emmanuel Levinas, *Totality and Infinity: An Essay on Exteriority,* trans. Alphonso Lingis (Boston: Kluwer Academic, 1988), 242–43.

5. Peter Stamatov, *The Origins of Global Humanitarianism: Religion, Empires, and Advocacy* (New York: Cambridge University Press, 2013); Kelly Oliver, *Carceral Humanitarianism: Logics of Refugee Detention* (Minneapolis: University of Minnesota Press, 2017), 47–48.

6. It is worth noting that this raises the underlying question of what types of pain and which subjects' pain are considered "necessary" or "legitimate." Bruno Cabanes, *The Great War and the Origins of Humanitarians, 1918–1924* (New York: Cambridge University Press, 2014); Lynn Hunt, *Inventing Human Rights: A History* (New York: W. W. Norton, 2008).

7. Several scholars read this as a critical development, such as Karen Halttunen, "Humanitarianism and the Pornography of Pain in Anglo-American Culture," *American Historical Review* 100, no. 2 (1995): 303–34, and Patricia A. Owens, "Xenophilia, Gender, and Sentimental Humanitarianism," *Alternatives: Global, Local, Political* 29 (2004): 285–304.

8. Emma Hutchison, *Affective Communities in World Politics: Collective Emotions after Trauma* (Cambridge: Cambridge University Press, 2016), 7–12.

9. In terms of philosophical debate, Levinas commits to an other that transcends, which is a problematic place for ethics, but this discussion goes beyond the scope of this text. David Campbell, "Why Fight?

Humanitarianism, Principles, and Poststructuralism," in *Ethics and International Relations,* ed. Hakan Seckinelgin and Hideaki Shinoda, 132–60 (New York: Palgrave Macmillan, 2001).

10. Sinja Graf, *The Humanity of Universal Crime: Inclusion, Inequality, and Intervention in International Political Thought* (New York: Oxford University Press, 2021), 8.

11. Jacques Derrida, *The Animal That Therefore I Am,* trans. David Wills (New York: Fordham University Press, 2008), 107–8.

12. Primo Levi, *The Drowned and the Saved* (New York: Vintage, 1989), 25.

13. Levinas, "Name of a Dog," 152, emphasis added.

14. Derrida, *Animal That Therefore I Am,* 106.

15. There is a large literature on Levinas's ambiguous relationship with animals. See Peter Atterton, "Levinas and Our Moral Responsibility toward Other Animals," *Inquiry: An Interdisciplinary Journal of Philosophy* 54, no. 6 (2011): 633–49; Peter Atterton, "Face to Face with the Other Animal?," in *Levinas and Buber: Dialogue and Difference,* ed. Peter Atterton, Matthew Claarco, and Maurice Friedman, 262–81 (Pittsburgh, Pa.: Duquesne University Press, 2004); David Clark, "On Being 'the Last Kantian in Nazi Germany': Dwelling with Animals after Levinas," in *Postmodernism and the Ethical Subject,* ed. Barbara Gabriel and Susan Ilcan, 41–84 (Montreal: McGill-Queen's University Press, 2004); John Llewelyn, "Am I Obsessed by Bobby? Humanism of the Other Animal?," in *Re-reading Levinas,* ed. Robert Bernasconi and Simon Critchley, 234–36 (Bloomington: Indiana University Press, 1991).

16. Immanuel Kant, *Anthropology from a Pragmatic Point of View,* trans. Robert B. Louden (New York: Cambridge University Press, 2006), 15, emphasis original.

17. Derrida, *Animal That Therefore I Am,* 99.

18. Brian Goodwin, *How the Leopard Changes Its Spots: The Evolution of Complexity* (Princeton, N.J.: Princeton University Press, 2001), 3–6; Dorian Sagan, *Cosmic Apprentice: Dispatches from the Edges of Science* (Minneapolis: University of Minnesota Press, 2013), 76–80.

19. Derrida, *Animal That Therefore I Am,* 24–28.

20. Michael Fagenblat, *A Covenant of Creatures: Levinas's Philosophy of Judaism* (Stanford, Calif.: Stanford University Press, 2010).

21. "Mission and Principles," Salvation Army Southern Territory, https://salvationarmypotomac.org/hrva/about-us/mission-principles/; "Fundamental Principles of the Red Cross and Red Crescent Movement," International Committee of the Red Cross, April 11, 2016, https://www.icrc.org/en/document/fundamental-principles-red-cross-and

-red-crescent; "Bearing Witness," Medicins sans frontiers/Doctors without Borders, https://www.doctorswithoutborders.org/who-we-are/principles/bearing-witness, emphasis added.

22. Pheng Cheah, *Inhuman Conditions: On Cosmopolitanism and Human Rights* (Cambridge, Mass.: Harvard University Press, 2006), 230–31.

23. Elizabeth Kolbert, *The Sixth Extinction: An Unnatural History* (New York: Picador, 2014).

24. Greg Goodale, *The Rhetorical Invention of Man: A History of Distinguishing Humans from Other Animals* (Lanham, Md.: Lexington Books, 2015). See also Joanna Bourke, *What It Means to Be Human: Historical Reflections from the 1800s to the Present* (Berkeley, Calif.: Counterpoint, 2011).

25. Michel Foucault, *The Order of Things: An Archaeology of the Human Sciences* (New York: Vintage, 1994); Makau Mutua, "Savages, Victims and Saviors: The Metaphor of Human Rights," *Harvard International Law Journal* 42, no. 1 (2001): 204–6.

26. Martha Finnemore, "Constructing Norms of Humanitarian Intervention," in *Conflict after the Cold War: Arguments on the Causes of War and Peace,* ed. Richard K. Betts, 262–79 (New York: Routledge, 2016).

27. Alexandre Lefebvre, *Human Rights and Care of the Self* (Durham, N.C.: Duke University Press, 2018), 4–7.

28. William E. Connolly, *Facing the Planetary: Entangled Humanism and the Politics of Swarming* (Durham, N.C.: Duke University Press, 2017), 8–12.

29. William J. McNeill, "Care for the Self: Originary Ethics in Heidegger and Foucault," *Philosophy Today* 42, no. 1 (1998): 53–55.

30. S. A. West, A. S. Griffin, and A. Gardner, "Social Semantics: Altruism, Cooperation, Mutualism, Strong Reciprocity and Group Selection," *European Society for Evolutionary Biology* 20 (2007): 415–32.

31. Michael Barnett, *Empire of Humanity: A History of Humanitarianism* (Ithaca, N.Y.: Cornell University Press, 2013), 17–18.

32. Nicholas Wheeler, *Saving Strangers: Humanitarian Intervention in International Society* (New York: Oxford International Press, 2002).

33. For the best reading of the origins of the R2P, see Anne Orford, *International Authority and the Responsibility to Protect* (Cambridge: Cambridge University Press, 2011).

34. Samuel Moyn, *Human Rights and the Uses of History* (New York: Verso, 2014).

35. Kendra Coutler, *Animals, Work, and the Promise of Interspecies Solidarity* (New York: Palgrave Macmillan, 2016), 1–2, 165–67; Charlotte E. Blattner, Kendra Coulter, and Will Kymlicka, "Introduction: Animal Labour and the Quest for Interspecies Justice," in *Animal Labour: A New*

Frontier of Interspecies Justice, ed. Charlotte E. Blattner, Kendra Coulter, and Will Kymlicka, 1–22 (New York: Oxford University Press, 2020).

36. See, e.g., "Thanks to Demining Dogs, Mozambique to Be Mine Free by 2014," Humanity and Inclusion, https://www.hi-us.org/thanks_to_demining_dogs_mozambique_to_be_mine_free_by_2014.

37. "Mine Detection Dogs," Marshall Legacy Institute, http://marshall-legacy.org/programs-2/mine-detection-dogs/.

38. Ron Verhagen, Christophe Cox, Robert Machangu, Bart Weetjens, and Mic Billet, "Preliminary Results on the Use of Cricetomys Rats as Indicators of Buried Explosives in Field Conditions," in *Mine Detection Dogs: Training, Operations, and Odour Detection,* ed. Ian G. McLean, 175–94 (Geneva: Geneva International Centre for Humanitarian Demining, 2003).

39. Sheila J. Bryant, "Pay It Forward: The Heifer International Story," *Journal of Agricultural and Food Information* 5, no. 3 (2003): 5–9.

40. For a brilliant, inspirational description of the materiality of humanitarian compounds, see Lisa Smirl, *Spaces of Aid: How Cars, Compounds and Hotels Shape Humanitarianism* (London: Zed Books, 2015).

41. Roland Paris, "The 'Responsibility to Protect' and the Structural Problems of Preventative Humanitarian Intervention," *International Peacekeeping* 21, no. 5 (2014): 569–603.

42. Mahmood Mamdani, *Saviors and Survivors: Darfur, Politics, and the War on Terror* (New York: Doubleday, 2009), 19–47.

43. David King, "The New Internationalists: World Vision and the Revival of American Evangelical Humanitarianism, 1950–2010," *Religions* 3, no. 4 (2012): 922–49.

44. Rafi Youatt, *Interspecies Politics: Nature, Borders, States* (Ann Arbor: University of Michigan Press, 2020); Delf Rothe, "Jellyfish Encounters: Science, Technology, and Security in the Anthropocene Ocean," *Critical Studies on Security* 8, no. 2 (2020): 145–59; John Dryzek and Jonathan Pickering, *The Politics of the Anthropocene* (New York: Oxford University Press, 2019); Anthony Burke, "Blue Screen Biosphere: The Absent Presence of Biodiversity in International Law," *International Political Sociology* 13, no. 3 (2019): 333–51; Audra Mitchell, "Beyond Biodiversity and Species: Problematizing Extinction," *Theory and Event* 33, no. 5 (2016): 23–42; Mark B. Salter, ed., *Making Things International 1: Circuits and Motion* (Minneapolis: University of Minnesota Press, 2015).

45. Anthony Burke, "Interspecies Cosmopolitanism: Non-human Power and the Grounds of World Order in the Anthropocene," *Review of International Studies* (April 12, 2022), https://doi.org/10.1017/S0260210522000171; Simon Dalby, *Anthropocene Geopolitics: Globalization, Security, Sustain-*

ability (Ottawa, Ont.: University of Ottawa Press, 2020); Madelein Fagan, "On the Dangers of an Anthropocene Epoch: Geological Time, Political Time and Post-human Politics," *Political Geography* 70 (2019): 55–63; Matthew Leep, "Stray Dogs, Post-humanism and Cosmopolitan Belongingness: Interspecies Hospitality in Times of War," *Millennium: Journal of International Studies* 47, no. 1 (2018): 45–66; Erika Cudworth and Stephen Hobden, *The Emancipatory Project of Posthumanism* (London: Routledge, 2017); Anthony Burke, Stefanie Fishel, Audra Mitchell, Simon Dalby, and Daniel J. Levine, "Planet Politics: A Manifesto for the End of IR," *Millennium: Journal of International Studies* 44, no. 3 (2016): 499–523; Veronique Pin-Fat, "Cosmopolitanism and the End of Humanity: A Grammatical Reading of Posthumanism," *International Political Sociology* 7, no. 3 (2013): 241–57.

46. Marcelo R. Sánchez-Villagra, *The Process of Animal Domestication* (Princeton, N.J.: Princeton University Press, 2022), 33.

47. See, e.g., Kenneth G. Furton and Lawrence J. Myers, "The Scientific Foundation and Efficacy of the Use of Canines as Chemical Dectors for Explosives," *Talanta* 54, no. 3 (2001): 487–500; Matthew Lewon, E. Kate Webb, Sydney M. Brotheridge, Christophe Cox, and Cynthia D. Fast, "Behavioral Skills Training in Scent Detection Research: Interactions between Trainer and Animal Behavior," *Journal of Applied Behavioral Analysis* 52, no. 3 (2019): 682–700.

48. At times I use the vocabulary of "companion species." However, this language is also occasionally used by humanitarian organizations that instrumentalize nonhuman animals. Donna Haraway, *The Companion Species Manifesto: Dogs, People, and Significant Otherness* (Chicago: Prickly Paradigm Press, 2003).

49. Max Weber, *Economy and Society,* ed. Guenther Roth and Claus Wittich (Berkeley: University of California Press, 1978), 24–26.

50. Derrida, *Animal That Therefore I Am,* 119–20; Neel Ahuja, *Biosecurities: Disease Intervention, Empire, and the Government of Species* (Durham, N.C.: Duke University Press, 2016), xi.

51. Sagan, *Cosmic Apprentice,* 80–86; Jane Bennett, *Vibrant Matter: A Political Ecology of Things* (Durham, N.C.: Duke University Press, 2009); Tim Morton, *Humankind: Solidarity with Nonhuman People* (New York: Verso, 2017); Samantha Frost, *Biocultural Creatures: Toward a New Theory of the Human* (Durham, N.C.: Duke University Press, 2016); Eugene Thacker, *After Life* (Chicago: University of Chicago Press, 2010).

52. Eric D. Schneider and Dorian Sagan, *Into the Cool: Energy Flow, Thermodynamics, and Life* (Chicago: University of Chicago Press, 2006), 143–45.

53. J. M. Coetzee, *The Lives of Animals,* ed. Amy Gutmann (Princeton, N.J.: Princeton University Press, 1999); Stefanie Fishel, *The Microbial State: Global Thriving and the Body Politic* (Minneapolis: University of Minnesota Press, 2016).

54. Dipesh Chakrabarty, "The Planet: An Emergent Humanist Category," *Critical Inquiry* 46, no. 1 (2019): 1–31.

55. Matthew Calarco, *Thinking through Animals* (Stanford, Calif.: Stanford University Press, 2015); Leyre Castro and Ed Wasserman, "Crows Understand Analogies," *Scientific American,* February 10, 2015, https://www.scientificamerican.com/article/crows-understand-analogies/.

56. Louise Barret, *Beyond the Brain: How Body and Environment Shape Animal and Human Minds* (Princeton, N.J.: Princeton University Press, 2011); Kristin Andrews, *The Animal Mind: The Philosophy of Animal Cognition* (New York: Routledge, 2014).

57. Marc Bekoff and Jessica Pierce, *Wild Justice: The Moral Lives of Animals* (Chicago: University of Chicago Press, 2010), 5.

58. Amanda Seed and Richard Byrne, "Animal Tool-Use," *Current Biology* 20, no. 23 (2010): 1032–39.

59. Aristotle, *Politics,* trans. C. D. C. Reeve (Indianapolis, Ind.: Hackett, 2017), 1253; Yevgeny Levkovich, "Legends of the Moscow Zoo: Reptilian Rumors and Killer Crocs," *Russia Beyond,* February 24, 2017, https://www.rbth.com/arts/2017/02/22/legends-of-the-moscow-zoo-reptilian-rumors-and-killer-crocs_707748; BBC News, "Famous Greek Riot Dog Loukanikos Dies," October 10, 2014, https://www.bbc.com/news/world-europe-29565725.

60. Giorgio Agamben, *The Open: Man and Animal,* trans. Kevin Attell (Stanford, Calif.: Stanford University Press, 2004), 26–28.

61. Agamben himself maintains that the distinction emerges from exposure to a certain set of potentialities bound up with the experience of a language, a perspective that this book does not endorse. Agamben, *Open,* 21–22.

62. Agamben, *Open,* 37.

63. Brian Massumi, *What Animals Teach Us about Politics* (Durham, N.C.: Duke University Press, 2014), 55–56; Alexander G. Weheliye, *Habeas Viscus: Racializing Assemblages, Biopolitics, and Black Feminist Theories of the Human* (Durham, N.C.: Duke University Press, 2014), 33–45.

64. Gregory Bateson, *Steps to an Ecology of Mind: Collected Essays in Anthropology, Psychiatry, Evolution, and Epistemology* (Chicago: University of Chicago Press, 2000), 209–38.

65. By point of comparison, human capacities for sociality, compassion, and

care did not develop a priori for the good of human welfare but as an evolutionary response to ecological constraints and random mutations. The implicit stance that virtues of care and concern must always already be oriented toward human life thus makes no sense even in the context of human evolution. This point illustrates the important political function of the distinction, to reify the division between nonhumans and humans, rather than its empirical or ethical function.

66. Anthropocentrism needs to be understood as more than a bias of humans toward a human-centered worldview. First, not all perspectives among those ascribed humanity share an agreement that the human constitutes a common starting point. Second, anthropocentrism discredits many liminal existences from being human, fully human, or on the cusp of humanity. Anthropocentrism is, in other words, a value system only from within one paradigm and is still capable of maintaining violent divisions within the human estate. Decolonial and anticolonial efforts are thus not inseparable from projects addressing nonhuman animal or species difference. Colonialism involves the promotion of a species, a way of thinking about the importance of species, and human difference, just as nonhuman animals function as signals of national identity, territorial integrity, and racial difference. Though the emphasis here is on nonhuman animal/species difference, both are key, intersecting dimensions. See Clair Jean Kim, *Dangerous Crossing: Race, Species, and Nature in a Multicultural Age* (Cambridge: Cambridge University Press, 2015); Kathryn Gillespie and Yamini Narayanan, "Animal Nationalisms: Multispecies Cultural Politics, Race, and the (Un)Making of the Settler Nation-State," *Journal of Intercultural Studies* 41, no. 1 (2020): 1–7; Billy-Ray Belcourt, "Animal Bodies, Colonial Subjects: (Re)locating Animality in Decolonial Thought," *Societies* 5, no. 1 (2015): 1–11.

67. Graf, *Humanity of Universal Crime,* 169–84; Kathryn Yusoff, *A Billion Black Anthropocenes or None* (Minnesota: University of Minnesota Press, 2019); Jairus Grove, *Savage Ecology: War and Geopolitics at the End of the World* (Durham, N.C.: Duke University Press, 2019), 35–58.

68. Scott Watson, "The 'Human' as Referent Object? Humanitarianism as Securitization," *Security Dialogue* 42, no. 1 (2011): 3–20; Makau Mutua, "Savages, Victims, and Saviors: The Metaphor of Human Rights," *Harvard International Law Journal* 42, no. 1 (2001): 201–9.

69. Michel Foucault, *The Order of Things: An Archaeology of the Human Sciences* (New York: Vintage, 1994). For more recent work, see Rosi Braidotti, *The Posthuman* (Malden, Mass.: Polity Press, 2013).

70. Ahmed Shaheed and Rose Parris Richter, "Is 'Human Rights' a Western Concept?," IPI Global Observatory, October 17, 2018, https://theglobal observatory.org/2018/10/are-human-rights-a-western-concept/.

71. Eyal Weizman, *The Least of All Possible Evils: Humanitarian Violence from Arendt to Gaza* (New York: Verso, 2012); Peter Redfield, "Vital Mobility and the Humanitarian Kit," in *Biosecurity Interventions: Global Health and Security in Question,* ed. Andrew Lakoff and Stephen J. Collier, 147–71 (New York: Columbia University Press, 2008).

72. Saidiya Hartman, *Scenes of Subjection: Terror, Slavery, and Self-Making in Nineteenth-Century America* (New York: Oxford University Press, 1997).

73. Sylvia Wynter, "Unsettling the Coloniality of Being/Power/Truth/ Freedom: Towards the Human, after Man, Its Overrepresentation — an Argument," *New Centennial Review* 3, no. 3 (2003): 257–337; Zakiyyah Iman Jackson, "New Directions in the Theorization of Race and Post-humanism," *Feminist Studies* 39, no. 3 (2013): 669–85.

74. Zakiyyah Iman Jackson, *Becoming Human: Matter and Meaning in an Antiblack World* (New York: New York University Press, 2020), 15–21.

75. Judith Butler, *Undoing Gender* (New York: Routledge, 2004), 13–16.

76. Barbara Schmitz, "'Something Else'? Cognitive Disability and the Human Form of Life," in *Disability and the Good Human Life,* ed. Jerome E. Bickenbach, Franziska Feider, and Barbara Schmitz, 52–57 (New York: Cambridge University Press, 2014).

77. Jackson argues that the liminality of Blackness exists on the threshold of human–animal distinction and constitutes it as plastic, subject to all manner of violence and violation, on this basis. Jackson, *Becoming Human,* 15–17.

78. Bénédicte Boisseron, *Afro-Dog: Blackness and the Animal Question* (New York: Columbia University Press, 2018), 37–80; Grégoire Chamayou, *Manhunts: A Philosophical History,* trans. Steven Rendall (Princeton, N.J.: Princeton University Press, 2012), 43–56.

79. Michael Ramirez, "'My Dog's Just Like Me': Dog Ownership as a Gender Display," *Symbolic Interaction* 29, no. 3 (2011): 373–91.

80. Haraway, *When Species Meet* (Minneapolis: University of Minnesota Press, 2008), 18.

81. For examples, see Peter Redfield, *Life in Crisis: The Ethical Journey of Doctors without Borders* (Berkeley: University of California Press, 2013), and Miriam I. Ticktin, *Casualties of Care: Immigration and the Politics of Humanitarianism* (Berkeley: University of California Press, 2011).

82. Didier Fassin, *Humanitarian Reason: A Moral History of the Present* (Berkeley: University of California Press, 2011), 26–29.

83. Giorgio Agamben, *Homo Sacer: Sovereign Power and Bare Life,* trans. Daniel Heller-Roazen (Stanford, Calif.: Stanford University Press, 1998), 6–8.

84. This is not to suggest that there are not differences between humans and other nonhuman animals, only that this system of thought cannot successfully articulate such differences. Frost, *Biocultural Creatures,* 22–24; Tim Morton, *Dark Ecology: For a Logic of Future Coexistence* (New York: Columbia University Press, 2016), 22.

85. Redfield, "Vital Mobility and the Humanitarian Kit," 164.

86. Smirl, *Spaces of Aid,* 7–8; Benjamin Meiches, "Nonhuman Humanitarians," *Review of International Studies* 45, no. 1 (2019): 1–19; Tom Scott-Smith, "The Fetishism of Humanitarian Objects and the Management of Malnutrition in Emergencies," *Third World Quarterly* 34, no. 5 (2013): 913–28; Polly Pallister-Wilkins, "Personal Protective Equipment in the Humanitarian Governance of Ebola: Between Individual Patient Care and Global Biosecurity," *Third World Quarterly* 37, no. 3 (2016): 507–23; Vicki Squire, "Desert 'Trash': Posthumanism, Border Struggles, and Humanitarian Politics," *Political Geography* 39 (2014): 11–21; Peter Redfield, "Fluid Technologies: The Bush Pump, the LifeStraw and Microworlds of Humanitarian Design," *Social Studies of Science* 46, no. 2 (2016): 159–83.

87. Slavoj Žižek, *The Universal Exception* (New York: Bloomsbury, 2014), 244–48.

88. Maria Puig de la Bellacasa, *Matters of Care: Speculative Ethics in More than Human Worlds* (Minneapolis: University of Minnesota Press, 2017); Lori Gruen, *Entangled Empathy: An Alternative Ethic for Our Relationships with Animals* (New York: Lantern Books, 2015).

89. William E. Connolly, *Identity/Difference: Democratic Negotiations of Political Paradox* (Minneapolis: University of Minnesota Press, 2002), xx, 92.

90. Sánchez-Villagra, *Process of Animal Domestication,* 36–66.

91. Anna Tsing, *The Mushroom at the End of the World: The Possibility of Life in Capitalist Ruins* (Princeton, N.J.: Princeton University Press, 2015), 235–55.

92. Jason Hribal and Jeffrey St. Clair, *Fear of the Animal Planet: The Hidden History of Animal Resistance* (N.p.: AK Press, 2011).

93. Massumi, *What Animals Teach Us,* 28–35.

94. Jenny Edkins, "Humanitarianism, Humanity, Human," *Journal of Human Rights* 2, no. 2 (2003): 253–85; Jennifer Hyndman, "Managing Difference: Gender and Culture in Humanitarian Emergencies," *Gender, Place, and*

Culture: A Journal of Feminist Geography 5, no. 3 (1998): 241–60; Debbie Lisle, "Humanitarian Travels: Ethical Communication in *Lonely Planet* Guidebooks," *Review of International Studies* 34, no. 1 (2008): 155–72; Chandra Mohanty, "Under Western Eyes: Feminist Scholarship and Colonial Discourses," *boundary 2* 12, no. 3 (1984): 333–58.

95. It is worth noting that anthropocentrism is a result of a particular historical-cultural context and not innate to a universal human perspective. The conditions for possibly structuring a being's mode of access, its perception of a reality, are, to a degree, plastic. There are thus multiple human modes of perception, just like there are multiple nonhuman modes of access. The point is that these filters, however singular, cannot be shed because they are constitutive of the thing in question. Perspective or mode of access is arguably what constitutes a thing in the first place.

96. Mark Estren, "The Neoteny Barrier: Seeking Respect for the Non-cute," *Journal of Animal Ethics* 2, no. 1 (2012): 6–11.

97. Morton, *Dark Ecology,* 108.

98. Haraway, *When Species Meet,* 52, emphasis original.

99. Liza Grauerholz, "Cute Enough to Eat: The Transformation of Animals into Meat for Human Consumption in Commercialized Images," *Humanity and Society* 31, no. 4 (2007): 334–54.

100. Liisa Malkki, *The Need to Help: The Domestic Arts of International Humanitarianism* (Durham, N.C.: Duke University Press, 2015), 15–18.

101. Michael Barnett and Thomas G. Weiss, *Humanitarianism in Question: Politics, Power, Ethics* (Ithaca, N.Y.: Cornell University Press), 235.

102. For a sampling of different popular proponents of humanitarianism despite its shortcomings, see Dexter Filkins, "The Moral Logic of Humanitarian Intervention," *New Yorker,* September 16, 2019, https://www .newyorker.com/magazine/2019/09/16/the-moral-logic-of-humanitarian -intervention; Gary Bass, *Freedom's Battle: The Origins of Humanitarian Intervention* (New York: Vintage, 2009); Michael Ignatieff, *The Lesser Evil: Political Ethics in an Age of Terror* (Princeton, N.J.: Princeton University Press, 2005); Scott Horton, "All the Missing Souls: Six Questions for David Scheffer," *Harper's,* February 7, 2012, https://www.global policy.org/international-justice/universal-jurisdiction-6-31/51269-all -the-missing-souls-six-questions-for-david-scheffer-.html; Robert Mardini, "In a Changing World, Humanitarianism Needs to Be Smarter and More Innovative than Ever," *The National,* December 16, 2019, https:// www.thenational.ae/opinion/comment/in-a-changing-world-humanitar ianism-needs-to-be-smarter-and-more-innovative-than-ever-1.952100.

103. Hunt, *Inventing Human Rights,* 8–14.

104. Kathryn Sikkink, *The Justice Cascade: How Human Rights Prosecutions Are Changing World Politics* (New York: W. W. Norton, 2011).

105. Margaret Keck and Kathryn Sikkink, *Activists beyond Borders: Advocacy Networks in International Politics* (Ithaca, N.Y.: Cornell University Press, 1998), 25–30.

106. Stephen Kloos, "Humanitarianism from Below: Sowa Rigpa, the Traditional Pharmaceutical Industry, and Global Health," *Medical Anthropology* 39, no. 2 (2020): 167–81; Lilie Chouliaraki, "The Theatricality of Humanitarianism: A Critique of Celebrity Advocacy," *Communication and Critical/Cultural Studies* 9, no. 1 (2012): 1–21.

107. Jean-Luc Nancy, "Ce que les peoples arabes nous signifient," *Libération*, March 28, 2011, https://www.liberation.fr/planete/2011/03/28/ce -que-les-peuples-arabes-nous-signifient_724744.

108. David Kennedy, *The Dark Side of Virtue: Reassessing International Humanitarianism* (Princeton, N.J.: Princeton University Press, 2005), 3–36, 342.

109. Kennedy, 348.

110. Samuel Moyn, *The Last Utopia: Human Rights in History* (Cambridge, Mass.: Belknap Press of Harvard University Press, 2012), 7–11, 212–19.

111. Jessica Whyte, *The Morals of the Market: Human Rights and the Rise of Neoliberalism* (New York: Verso, 2019).

112. Costas Douzinas, *Human Rights and Empire: The Political Philosophy of Cosmopolitanism* (New York: Routledge, 2007).

113. Stephen Hopgood, *The Endtimes of Human Rights* (Ithaca, N.Y.: Cornell University Press, 2013), 166–70.

114. Sharon Sliwinski, "The Aesthetics of Human Rights," *Culture, Theory, and Critique* 50, no. 1 (2009): 23–39.

115. Malkki, *Need to Help,* 17–18.

116. Gilles Deleuze and Félix Guattari, *A Thousand Plateaus: Capitalism and Schizophrenia,* trans. Brian Massumi (Minneapolis: University of Minnesota Press, 1987), 424–28. Deleuze and Guattari offer a much more precise application of this concept in relation to the genesis of the state apparatus. Obviously, this is a different context, and the concept is employed to map a different rhizome.

117. Deleuze and Guattari, 437–48.

118. Deleuze and Guattari, 464–66; Manuel DeLanda, *A Thousand Years of Nonlinear History* (New York: Zone Books, 1997), 24–27.

119. Antoine Traisnel argues that a sociotechnical shift occurs in modernity that frames the animal in new terms, capturing it as an epistemological object in a way that differs from the agonistic relation of hunting. Traisnel

uses capture in a slightly different valence than Deleuze and Guattari, but the epistemological break he cites is a crucial precondition of humanitarian uses of anthropocentric reason. Traisnel, *Capture: American Pursuits and the Making of a New Animal Condition* (Minneapolis: University of Minnesota Press, 2020), 1–25.

120. Deleuze and Guattari, *A Thousand Plateaus,* 425.

121. See, e.g., Nicola Perugini and Neve Gordon, *The Human Right to Dominate* (New York: Oxford University Press, 2015); Randall Williams, *The Divided World: Human Rights and Its Violence* (Minneapolis: University of Minnesota Press, 2010). In contrast, see Joe Hoover, *Reconstructing Human Rights: A Pragmatist and Pluralist Inquiry in Global Ethics* (New York: Oxford University Press, 2016).

122. John Protevi, *Edges of the State* (Minneapolis: University of Minnesota Press, 2019), 56–69.

123. Bourke, *What It Means to Be Human,* 69–73.

124. Tom Regan, *The Case for Animal Rights* (Berkeley: University of California Press, 2004), 294–97.

125. Lauren Berlant, *Cruel Optimism* (Durham, N.C.: Duke University Press, 2011), 23–25.

126. Slavoj Žižek, *The Sublime Object of Ideology* (New York: Verso, 2009), 74–75.

127. The genre of animal rights literature illustrates the connection. Animal rights advocates make overtly rational explanations of the legitimacy of animal rights by comparing nonhuman animal abilities for cognition with those of the rights-bearing subject. However, equally important is the emphasis on a nonhuman animal's capacity to experience suffering as well as the apparent arbitrariness of this suffering for the animal. This image works at the level of sympathy and resonates with humanitarianism's understanding of the pain of others.

128. Bourke, *What It Means to Be Human,* 69–73.

129. Lauren Berlant, *Desire/Love* (Brooklyn, N.Y.: Punctum Books, 2012), 20.

130. Berlant, 108.

131. Berlant, 75.

132. Weizman, *Least of All Possible Evils,* 7–9.

133. Danielle Celermajer, David Schlosberg, Lauren Rickards, Makere Stewart-Harawira, Mathias Thaler, Petra Tschakert, Blanche Verlie, and Christine Winter, "Multispecies Justice: Theorists, Challenges, and a Research Agenda for Environmental Politics," *Environmental Politics* 30, no. 1–2 (2021): 119–40.

134. Tore Fougner, "Engaging the 'Animal Question' in International Relations," *International Studies Review* 23, no. 3 (2021): 862–86.

135. Manuel DeLanda, *Intensive Science and Virtual Philosophy* (New York: Bloomsbury, 2013), 75.

136. Deleuze and Guattari, *A Thousand Plateaus*, 307–99.

137. Thom van Dooren, *The Wake of Crows: Living and Dying in Shared Worlds* (New York: Columbia University Press, 2019), 73–88; Claire Colebrook, *Death of the Posthuman* (Ann Arbor, Mich.: Open Humanities Press, 2014), 1:142–44.

138. Grégoire Chamayou, *Manhunts: A Philosophical History,* trans. Steven Randall (Princeton, N.J.: Princeton University Press, 2012), 16–18.

139. "HeroRATs," APOPO, 2022, https://apopo.org/herorats/?v=7516fd 43adaa.

140. Coco McCabe, "Small Herds Support Families during Hard Times," *Oxfam Impact Reports,* February 22, 2010, 1–2.

1. DOGS AND THE POLITICS OF DETECTING EXPLOSIVES

1. Ming Mei, "Cambodians Hold Religious Ceremony for Death of Landmine Detection Dog," *Asia and Pacific,* May 26, 2019, http://www .xinhuanet.com/english/2019-05/26/c_138090707.htm.

2. "International Campaign to Ban Landmines," *The Monitor,* November 20, 2018, http://the-monitor.org/media/2918822/PressRelease_ LandmineMonitor2018_embargoed_revised.pdf.

3. https://cmac.gov.kh/.

4. Mei, "Cambodians Hold Religious Ceremony."

5. Shira Li Bartov, "Bomb Detection Dog Is Honored with Hero's Funeral in Viral Video," *Newsweek,* March 8, 2022, https://www.newsweek.com/ bomb-detection-dog-honored-heros-funeral-viral-video-1685948.

6. "Cambodia's Mine Dog Heroes Honoured," Eleven Media Group, June 7, 2019, https://elevenmyanmar.com/news/cambodias-mine-dog -heroes-honoured-asianewsnetwork.

7. Marissa Carruthers, "Retired Landmine Detection Dogs Find a Home at Cambodian Charity after Working in 'Some of the Most Horrific Parts of the World,'" *South China Morning Post,* February 21, 2018, https:// www.scmp.com/lifestyle/article/2133973/retired-landmine-detection -dogs-find-home-cambodian-charity-after-working.

8. John Blewitt, "What's New Pussycat? A Genealogy of Animal Celebrity," *Celebrity Ecologies* 4, no. 2 (2019): 326–30.

9. For an analysis of the redefinition of family to include dogs, see Andrea Laurent-Simpson, *Just Like Family* (New York: New York University Press, 2021).

10. Haraway, *Companion Species Manifesto*.

11. Alexandra Horowitz, *Inside of a Dog: What Dogs See, Smell, and Know* (New York: Scribner, 2009).

12. Brian Hare and Vanessa Woods, "We Didn't Domesticate Dogs. They Domesticated Us," *National Geographic*, March 3, 2013, https://www.nationalgeographic.com/news/2013/3/130302-dog-domestic-evolution-science-wolf-wolves-human/; Brian Hare and Vanessa Woods, *The Genius of Dogs: How Dogs Are Smarter than You Think* (New York: Plume, 2013).

13. Sánchez-Villagra, *Process of Animal Domestication*, 36–40; Cat Warren, *What the Dog Knows: Scent, Science and the Amazing Ways Dogs Perceive the World* (New York: Touchstone, 2015).

14. Erika Cudworth and Steve Hobden, "The Posthuman Way of War," *Security Dialogue* 46, no. 6 (2015): 517–22; Tyler D. Parry and Charlton W. Yingling, "Slave Hounds and Abolition in the Americas," *Past and Present* 246, no. 1 (2020): 69–108; Boisseron, *Afro-dog*, 36–80; Sara E. Johnson, "'You Should Give Them Blacks to Eat': Waging Inter-American Wars of Torture and Terror," *American Quarterly* 61, no. 1 (2009): 65–92.

15. Having been the subject of two attacks by three dogs in the past decade, my anecdotal observation is that canine–human interactions are always political encounters mediated by race, class, and gender relations. Susan Hunter and Richard A. Brisbin Jr., *Pet Politics: The Political and Legal Lives of Cats, Dogs, and Horses in Canada and the United States* (West Layfatte, Ind.: Purdue University Press, 2016), 207–48.

16. Claire Jean Kim, *Dangerous Crossing: Race, Species, and Nature in a Multicultural Age* (Cambridge: Cambridge University Press, 2015), 255–89.

17. I am grateful to Derek Denman for bringing Loukanikos to my attention and for many vivacious conversations about dogs and politics. Graeme Wearden, "Greece's Riot Dog Loukanikos Dies," *The Guardian*, October 9, 2014, https://www.theguardian.com/business/2014/oct/09/greece-riot-dog-loukanikos-dies-eurozone-crisis.

18. Merrily Weisbord and Kim Kachanoff, *Dogs with Jobs: Working Dogs around the World* (Chicago: University of Chicago Press, 2012), 54–56.

19. Katie Rogers, "Trump Says 'Beautiful' and 'Talented' Dog Injured in al-Baghdadi Raid," *New York Times*, October 27, 2019, https://www.nytimes.com/2019/10/27/us/politics/trump-dog-al-baghdadi-raid.html.

20. Rachel Treisman, "Patron the Bomb-Sniffing Dog Cements His Hero Status with a Presidential Medal," NPR, May 9, 2022, https://www.npr.org/2022/05/09/1097585032/patron-dog-ukraine-zelenskyy-medal.

21. "Animal Cruelty Facts and Stats," Humane Society of the United States, 2019, https://www.humanesociety.org/resources/animal-cruelty-facts-and-stats.

22. Sue Donaldson and Will Kymlicka, *Zoopolis: A Political Theory of Animal Rights* (New York: Oxford University Press, 2013), 141.

23. It is important to note that there is no such thing as uniform animal interests. The biopolitics of industrial meat production arguably benefits human companion animals, such as dogs and cats, even as it harms other species, such as chickens and cows. Humans also have a mixed relationship with these institutions, which often promote violence against human laborers alongside nonhuman animals. Timothy Pachirat, *Every Twelve Seconds: Industrialized Slaughter and the Politics of Sight* (New Haven, Conn.: Yale University Press, 2013), 275.

24. Carol J. Adams, *The Sexual Politics of Meat: A Feminist-Vegetarian Critical Theory,* Anniversary ed. (New York: Bloomsbury Academic, 2015), 28.

25. Cary Wolfe, *Before the Law: Humans and Other Animals in a Biopolitical Frame* (Chicago: University of Chicago Press, 2012), 26–27.

26. Grégoire Chamayou, *Manhunts: A Philosophical History,* trans. Steven Rendall (Princeton, N.J.: Princeton University Press, 2012), 3–4.

27. Ryan Hediger, "Dogs of War: The Biopolitics of Loving and Leaving the U.S. Canine Forces in Vietnam," *Animal Studies Journal* 2, no. 1 (2013): 55–73; Janet M. Alger and Steven F. Alger, "Canine Soliders, Mascots, and Stray Dogs in U.S. Wars: Ethical Considerations," in *Animals in War: Studies of Europe and North America,* ed. Ryan Hediger, 77–104 (Leiden, Netherlands: Brill, 2013).

28. For an excellent account of the development of demining dogs in the British context, see Robert G. W. Kirk, "In Dogs We Trust? Intersubjectivity, Response-Able Relations, and the Making of Mine Detector Dogs," *Journal of Historical Behavioral Science* 50, no. 1 (2014): 1–36.

29. Seth Paltzer, "The Dogs of War: The U.S. Army's Use of Canines in WWII," National Museum of the United States Army, June 2, 2016, https://army history.org/the-dogs-of-war-the-u-s-armys-use-of-canines-in-wwii/.

30. For a rich reading of great power armed conflict as implicit in a form of race war, see Alexander Barder, *Global Race War: International Politics and Racial Hierarchy* (New York: Oxford University Press, 2021), 135–60.

31. Parry and Yingling, "Slave Hounds and Abolition in the Americas," 75–80.

32. Parry and Yingling, 75.
33. Fiona Allon and Lindsay Barrett, "That Dog Was Marine! Human–Dog Assemblages in the Pacific War," *Animal Studies Journal* 4, no. 1 (2015): 126–47.
34. Kirk, "In Dogs We Trust?," 1–36.
35. "Mine Detection Dogs"; Dan Hayner, "The Evolution of Mine Detection Dog Training," *Journal of Conventional Weapons Destruction* 7, no. 1 (2003): 72.
36. United Nations Mine Action Service, *Guide to Occupational Health and General Dog Care,* International Mine Action Standard (New York: United Nations Mine Action Service, March 2008), 1–9.
37. CMAC, for instance, boasts a program that involves visiting dogs.
38. Eyal Weizman, *The Least of All Possible Evils: Humanitarian Violence from Arendt to Gaza* (London: Verso, 2012), 6–17.
39. Ann Laura Stoler, *Imperial Debris: On Ruins and Ruination* (Durham, N.C.: Duke University Press, 2013), 3–4, 54–55.
40. Rebecca J. Sargisson, Ian G. McLean, Dr. Jennifer Brown, and Håvard Bach, "Environmental Determinants of Landmine Detection by Dogs: Findings from a Large-Scale Study in Afghanistan," *Research and Development: The Journal of ERW and Mine Action* 16, no. 2 (2012): 74–80.
41. Warren, *What the Dog Knows.*
42. Irit Gazit and Joseph Terkel, "Domination of Olfaction over Vision in Explosives Detection by Dogs," *Applied Animal Behavioral Science* 82, no. 1 (2003): 65–73.
43. Peter Tyson, "Dogs' Dazzling Sense of Smell," PBS, October 3, 2012, https://www.pbs.org/wgbh/nova/article/dogs-sense-of-smell/.
44. Sargisson et al., "Environmental Determinants."
45. Maki Habib, "Mine Clearance Techniques and Technologies for Effective Humanitarian Demining," *Journal of Conventional Weapons Destruction* 6, no. 1 (2002): 63.
46. "Mine Detection Dogs."
47. Antoine Bousquet, *The Eye of War: Military Perception from the Telescope to the Drone* (Minneapolis: University of Minnesota Press, 2018), 7–8.
48. Leah Zani, "Bomb Ecologies," *Environmental Humanities* 10, no. 2 (2018): 528–31.
49. Anthony A. Faust, C. J. de Ruiter, Anneli Ehlerding, John E. McFee, Eirik Svinsås, and Arthur D. van Rheenen, "Observations on Military Exploitation of Explosives Detection Technologies," *Proceedings of the SPIE* 8017 (2011).

50. On the notion of biological assemblages, see DeLanda, *A Thousand Years of Nonlinear History*, 103–48.

51. Erica N. Feubacher and C. D. L. Wynne, "A History of Dogs as Subjects in North American Experimental Psychological Research," *Comparative Cognition and Behavior Reviews* 6, no. 46 (2011): 46–71.

52. DeLanda, *Intensive Science and Virtual Philosophy*, x. The question of the evolution of dogs is open and shifting, especially regarding their coevolutionary involvement with humans. Rachael Lallensack, "Ancient Genomes Heat Up Dog Domestication Debate," *Nature*, July 18, 2017, https://www.nature.com/news/ancient-genomes-heat-up-dog-domestication-debate-1.22320.

53. Thomas Nagel, "What Is It Like to Be a Bat?," *Philosophical Review* 83, no. 4 (1974): 435–50.

54. Jakob von Uexküll, *A Foray into the Worlds of Animals and Humans: With a Theory of Meaning*, trans. Joseph D. O'Neil (Minneapolis: University of Minnesota Press, 2010), 51–52.

55. Eleana J. Kim, "Toward an Anthropology of Landmines: Rogue Infrastructure and Military Waste in the Korean DMZ," *Cultural Anthropology* 31, no. 2 (2016): 162–87; David Henig, "Iron in the Soil: Living with Military Waste in Bosnia-Herzegovina," *Anthropology Today* 28, no. 1 (2002): 21–23; Rob Nixon, "Of Land Mines and Cluster Bombs," *Cultural Critique* 67 (2007): 160–74; Jon D. Unruh, Nikolas C. Heynen, and Peter Hossler, "The Political Ecology of Recovery from Armed Conflict: The Case of Landmines in Mozambique," *Political Geography* 22, no. 8 (2003): 841–61; Ann Laura Stoler, "Imperial Debris: Reflections on Ruin and Ruination," *Cultural Anthropology* 23, no. 2 (2008): 191–219.

56. Graham Harman, *Guerrilla Metaphysics* (Chicago: Open Court, 2005), 251.

57. Morton, *Dark Ecology*, 17–18.

58. Alfred North Whitehead, *Science and the Modern World* (New York: Simon and Schuster, 1997), 52.

59. Morton, *Dark Ecology*, 7–11.

60. Rob Nixon, *Slow Violence and the Environmentalism of the Poor* (Cambridge, Mass.: Harvard University Press, 2011), 7.

61. Kim, "Toward an Anthropology of Landmines," 166–69.

62. Grove, *Savage Ecology*, 4.

63. Leah Zani, *Bomb Children: Life in the Former Battlefields of Laos* (Durham, N.C.: Duke University Press, 2019), 112–14.

64. Morton, *Dark Ecology,* 21.

65. Stoler, *Imperial Debris,* 130–35.

66. Mary Kaldor, *New Wars, Old Wars: Organized Violence in a Global Era* (Malden, Mass.: Polity Press, 2008).

67. Tarak Barkawi, *Globalization and War* (Lantham, Md.: Rowman and Littlefield, 2005).

68. Sven Lindqvist, *A History of Bombing* (New York: New Press, 2003).

69. Agamben, *Homo Sacer,* 7.

70. Nixon, *Slow Violence.*

71. International law surrounding crimes against humanity and war crimes often seeks to counteract this model of moral intention by broadly holding perpetrators of military violence accountable for a range of actions they supported regardless of the specific victims.

72. Bennett, *Vibrant Matter,* 36.

73. Graham Harman, *Circus Philosophicus* (Washington, D.C.: Zero Books, 2010), 49–50.

74. Whether it is politically important is an entirely different question. All political statements involve paraphrase.

75. In a sense, one of the more radical dimensions of humanitarian politics is a willingness to take responsibility for acts of violence that are a result, not of one's own action, but of an indeterminate collective. The decision to take responsibility involves some recognition of vicarious causality because inaction, ignoring a problem like land mines, would also contribute to the continuation of the status quo, after epistemic awareness of the situation could have changed a potential future. Humanitarianism is thus also more complex in its articulation of responsibility than many classical principles or articulations of it allow for.

76. This argument should not be read as disputing the need to condemn the act of laying mines or other unexploded ordnance. Rather, it highlights the role of fiction in the formation of legal responsibility. The fictional status of concepts of like intent, however, does not deprive them of political importance as instruments for reducing the violence of mines or holding parties accountable for promoting violent conditions. It is possible simultaneously to support international laws of armed conflict, argue for their expansion, and criticize their conception of the dynamics that produce violence.

77. Bousquet, *Eye of War,* 22.

78. Greg Siegel, *Forensic Media: Reconstructing Accidents in Accelerated Modernity* (Durham, N.C.: Duke University Press, 2014), 15–25.

79. Siegel, 27.

80. Graham Harman, *Weird Realism: Lovecraft and Philosophy* (Washington, D.C.: Zero Books, 2012), 40–42.

81. Adrianna Cavarero, *Horrorism: Naming Contemporary Violence* (New York: Columbia University Press, 2009), 9.

82. *Play* here is a term describing the ambiguities of contingency. See Georges Bataille, "On the Ambiguity of Pleasure and Play," *Theory, Culture, and Society* 35, no. 4–5 (2018): 236–37.

83. I inherit the term from Michael E. Gardiner, "Wild Publics and Grotesque Symposiums: Habermas and Bakhtin on Dialogue, Everyday Life, and the Public Sphere," *Sociological Review* 52, no. 1 (2004): 28–48.

84. I will not go into this point in detail, but obviously nonhuman animals also trigger these mines in ecology, and so the victims are not only human communities. Achilles Mbembe, "Necropolitics," *Public Culture* 15, no. 1 (2003): 11–40.

85. Diana Pardo Pedraza, "On Landmines and Suspicion: How (Not) to Walk Explosive Fields," online forum, *Environment and Planning D: Society and Space,* March 9, 2020, https://www.societyandspace.org/articles/on-landmines-and-suspicion-how-not-to-walk-explosive-fields; Kirk, "In Dogs We Trust?"; McLean, *Mine Detection Dogs*; Tycie Horsley, "Child-to-Child Risk Education," *Journal of Conventional Weapons Destruction* 19, no. 2 (2015): 32–34.

86. Vicki Hearne, *Animal Happiness: A Moving Exploration of Animals and Their Emotions* (New York: Skyhorse, 2007), 229.

87. Stewart Hilliard, "Principles of Animal Learning," in McLean, *Mine Detection Dogs,* 28–35.

88. Baruch Spinoza, *The Ethics, Treatise on the Emendation of the Intellect, and Selected Letters,* trans. Samuel Shirley (New York: Hackett, 1992).

89. James O'Heare, "Explaining and Changing People's Use of Aversive Stimulation in Companion Animal Training," *Journal of Applied Companion Animal Behavior* 1, no. 1 (2007): 15.

90. On desire as a productive faculty, see Daniel W. Smith, "Deleuze and the Question of Desire: Toward an Immanent Theory of Ethics," *Parrhesia* 2 (2007): 66–78.

91. Maria Hynes and Scott Sharpe, "Affected with Joy: Evaluating the Mass Actions of the Anti-globalisation Movement," *Borderlands: e-journal* 8, no. 3 (2009): 1–21.

92. Gilles Deleuze and Félix Guattari, *Anti-Oedipus: Capitalism and Schizophrenia* (Minneapolis: University of Minnesota Press, 1983), 29.

93. Gino Strada, "The Horror of Land Mines," *Scientific American* 274, no. 5 (1996): 40–45.

94. Cengiz Erisen, Milton Lodge, and Charles S. Taber, "Affective Contagion in Effortful Political Thinking," *Political Psychology* 35, no. 2 (2014): 187–206; Elisabetta Palagi, Velia Nicotra, and Giada Cordoni, "Rapid Mimicry and Emotional Contagion in Domestic Dogs," *Royal Society Open Science* 2, no. 12 (2015): 150505.

95. Vinciane Despret, "Responding Bodies and Partial Affinities in Human–Animal Worlds," *Theory, Culture, and Society* 30, no. 7–8 (2013): 62. See also Despret, "The Body We Care For: Figures of Anthropo-Zoo-Genesis," *Body and Society* 10, no. 2–3 (2004): 131.

96. Ben Anderson, *Encountering Affect: Capacities, Apparatuses, Conditions* (Burlington, Vt.: Ashgate, 2014), 8–14.

97. Massumi, *What Animals Teach Us,* 93–94.

98. Annett Schirmer, Cui Shan Seow, and Traevor B. Penney, "Humans Process Dog and Human Facial Affect in Similar Ways," *PLoS ONE* 8, no. 9 (2013): e0074591.

99. Pedraza, "On Landmines and Suspicion."

100. Watson, "The 'Human' as Referent Object?"

101. Morton, *Dark Ecology,* 114–28.

2. HEROES, RATS, AND THE PREDICAMENT OF JUSTICE

1. "HeroRAT Magawa—PDSA Gold Medal," People's Dispensary for Sick Animals, September 2020, https://www.pdsa.org.uk/what-we-do/animal-awards-programme/pdsa-gold-medal.

2. "HeroRAT Magawa."

3. "Magawa the Mine-Detecting Rat Wins PDSA Gold Medal," BBC News, September 24, 2020, https://www.bbc.com/news/world-54284952; Anna Schaverien, "Rat That Sniffs Out Land Mines Receives Award for Bravery," *New York Times,* September 25, 2020, https://www.nytimes.com/2020/09/25/world/europe/giant-rat-magawa-medal.html; Jessie Yeung, "'Hero Rat' Wins Gold Medal from UK Charity for Hunting Landmines," CNN, September 25, 2020, https://www.cnn.com/2020/09/25/asia/hero-rat-landmines-award-intl-hnk-scli/index.html.

4. Schaverien, "Rat That Sniffs Out Land Mines."

5. Chad De Guzman, "Magawa the Landmine-Sniffing Rat Was an Inter-

national Hero. His Work Is Far from Over," *Time,* January 13, 2022, https://time.com/6138994/magawa-dies-landmines-cambodia/.

6. Michael McCormick, "Rats, Communications, and Plague: Toward an Ecological History," *Journal of Interdisciplinary History* 34, no. 1 (2003): 1–25; William H. McNeill, *Plagues and Peoples* (New York: Doubleday, 1997), 138–39; Anne Karin Hufthammer and Lars Walloe, "Rats Cannot Have Been Intermediate Hosts for *Yersinia pestis* during Medieval Plague Epidemics in Northern Europe," *Journal of Archaeological Science* 40, no. 4 (2013): 1752–59.

7. Yvonne Baskin, *A Plague of Rats and Rubber-vines: The Growing Threat of Species Invasions* (Washington, D.C.: Island Press, 2002).

8. Nicholas H. A. Evans, "Blaming the Rat? Accounting for Plague in Colonial India," *Medical Anthropology Theory* 5, no. 3 (2018): 15–42; Mary Douglas, *Purity and Danger: An Analysis of Concepts of Pollution and Taboo* (New York: Routledge, 1966), 72.

9. Aristotle, *History of Animals,* trans. D'Arcy Wentworth Thompson, book VI, chap. 37, http://classics.mit.edu/Aristotle/history_anim.html.

10. Robin Hutton, "The Surprising Story of the Only Cat Ever to Win the Highest Honor for Animal Military Gallantry," *Time,* September 18, 2018, https://time.com/5396568/simon-cat-war-medal/.

11. Animus toward rats depends on historical, geographic, and cosmological context, but dread of rats remains a recurrent feature across multiple social contexts. Adrian Daub, "All Hail the Rat King: From Martin Luther to *The Nutcracker,* Germany's Original National Nightmare Was a Tangled Knot of Writhing Rats," *Longreads,* December 2019, https://longreads.com/2019/12/11/rat-king-germany-nutcracker/.

12. Frederick J. Wertz, "The Rat in Psychological Science," *Humanistic Psychologist* 28, no. 1–3 (2000): 89.

13. Edmund Ramsden, "From Rodent Utopia to Urban Hell: Population, Pathology, and the Crowded Rats of NIMH," *ISIS: Journal of the History of Science Society* 102, no. 4 (2011): 659–88.

14. Joanna Makowska and Daniel M. Weary, "Assessing the Emotions of Laboratory Rats," *Applied Animal Behavior Science* 148 (2013): 112.

15. Note the anthropocentric standard of viewing complex social behavior as human rather than humans iterating a broader nonhuman process of complex socialization. Birgitta Edelman, "Rats Are People Too! Rat–Human Relations Re-rated," *Anthropology Today* 18, no. 3 (2002): 8.

16. Many of these studies never discuss the underlying assumption that these are recognizable as intrinsically human traits. University of Georgia, "Rats

Capable of Reflection on Mental Processes," *Science Daily,* March 9, 2007, http://www.sciencedaily.com/releases/2007/03/070308121856 .htm.

17. Shea O'Neill, "Scurrying alongside Humanity: A Co-evolutionary History of Rats and Humans," *History Matters,* https://historymatters .appstate.edu/sites/historymatters.appstate.edu/files/ratsandhumans .pdf.

18. Elizabeth Howell, "Rats in Spaaace! NASA Wants to Put Rodents on Space Station," August 19, 2014, https://www.space.com/26877-rats-in -space-nasa-experiment.html.

19. Michel Serres, *The Parasite,* trans. Lawrence R. Schehr (Minneapolis: University of Minnesota Press, 2007), 7–12.

20. "Our History," APOPO, 2017, https://www.apopo.org/en/who-we-are/ our-history.

21. Daniel Heller-Roazen, *The Enemy of All: Piracy and the Law of Nations* (New York: Zone Books, 2009), 155.

22. Didier Fassin, "Humanitarianism as a Politics of Life," *Public Culture* 19, no. 3 (2007): 499–520.

23. William E. Connolly, *A World of Becoming* (Durham, N.C.: Duke University Press, 2011), 23.

24. Bart Weetjens, "How I Taught Rats to Sniff Out Land Mines," TEDx Rotterdam, June 2010, https://www.ted.com/talks/bart_weetjens_how _i_taught_rats_to_sniff_out_land_mines.

25. "Mine Detection Rats: Efficiency and Effectiveness Study Using MDR Capability," Geneva International Centre for Humanitarian Demining, June 2016, 4, https://www.gichd.org/fileadmin/GICHD-resources/ rec-documents/APOPO-GICHD-Mine-Detection-Rats-30Jun2016.pdf.

26. "Mine Detection Rats," 6–10.

27. Kate B. La Londe, Amanda Mahoney, Timothy L. Edwards, Christophe Cox, Bart Weetjens, Amy Durgin, and Alan Poling, "Training Pouched Rats to Find People," *Journal of Applied Behavioral Analysis* 48, no. 1 (2015): 1–10; Amanda Mahoney, Timothy L. Edwards, Kate LaLonde, Negussie Beyene, Christophe Cox, Bart J. Weetjens, and Alan Poling, "Pouched Rats' *(Cricetomys gambianus)* Detection of *Salmonella* in Horse Feces," *Journal of Veterinary Behavior* 9, no. 3 (2014): 124–26; Amanda Mahoney, Kate La Londe, Timothy L. Edwards, Christophe Cox, Bart J. Weetjens, and Alan Poling, "Detection of Cigarettes and Other Tobacco Products by Giant African Pouched Rats *(Cricetomys gambianus),*" *Journal of Veterinary Behavior* 9, no. 5 (2014): 262–68.

28. "Mine Detection Rats," 4–5.

29. "Adopt an Animal," APOPO, 2020, https://www.apopo.org/en/adopt.

30. Malkki, *Need to Help,* 4–5.

31. Mamdani, *Saviors and Survivors*; Anne Orford, *Reading Humanitarian Intervention: Human Rights and the Use of Force in International Law* (New York: Cambridge University Press, 2008); Mohanty, "Under Western Eyes."

32. Weetjens, "How I Taught Rats."

33. Kelly Oliver, *Technologies of Life and Death: From Cloning to Capital Punishment* (New York: Fordham University Press, 2013), 36.

34. Bruno Latour, *Reassembling the Social: An Introduction to Actor-Network-Theory* (New York: Oxford University Press, 2005), 165–75.

35. I am grateful to Chad Shomura for helping me to articulate this insight.

36. APOPO also trains HeroDOGs for this work. Angela R. Freeman, Alexander G. Ophir, and Michael J. Sheehan, "The Giant Pouched Rat *(Cricetomys ansorgei)* Olfactory Receptor Repertoire," *PLoS ONE* 15, no. 4 (2020): e0221981.

37. Freeman et al. note that the genetic differences between *rattus* and *cricetomys* reveal that broad categories, such as "rat," organize social responses to ecological difference based on anthropocentric perceptions and prejudices.

38. Luisa Fernando Mendez Pardo and Andres M. Perez-Acosta, "Research in Colombia on Explosives Detection by Rats," *Journal of Conventional Weapons Destruction* 13, no. 3 (2009): 45–46. It is also worth noting that the category of species is itself porous, a reflection of preconceptions about identity. Sánchez-Villagra, *Process of Animal Domestication,* 73; Sagan, *Cosmic Apprentice,* 114.

39. Duke University, "Scientists Can Now See Sense of Smell," *Science-Daily,* July 1999, http://www.sciencedaily.com/releases/1999/07/990722064252.htm.

40. Markus Meister and Tobias Bonhoeffer, "Tuning and Topography in an Odor Map on the Rat Olfactory Bulb," *Journal of Neuroscience* 21, no. 4 (2001): 1351–53.

41. Oscar L. Vaccarezza, Liliana N. Sepich, and Juan H. Tamezzani, "The Vomeronasal Organ of the Rat," *Journal of Anatomy* 132, no. 2 (1981): 167–85.

42. "Pups to Mothers," APOPO, August 4, 2016, https://www.apopo.org/en/latest/2016/8/04/pups-to-heroes.

43. "MDR Training," APOPO, 2022, https://www.apopo.org/en/what-we-do/detecting-landmines-and-explosives/how-we-do-it/mdr-training.

44. "FAQs—Mine Action," APOPO, https://www.apopo.org/en/what -we-do/detecting-landmines-and-explosives/faqs.

45. Mahoney et al., "Detection of Cigarettes."

46. Mahoney et al., "Pouched Rats' *(Cricetomys gambianus)* Detection."

47. La Londe et al., "Training Pouched Rats."

48. "Animal Welfare," APOPO, https://www.apopo.org/en/herorats/animal -welfare.

49. "Animal Welfare."

50. Malkki, *Need to Help,* 79.

51. Richard P. Haynes, *Animal Welfare: Competing Conceptions and Their Ethical Implications* (New York: Springer, 2010), 7–12.

52. Jackson, *Becoming Human,* 14.

53. Vinciane Despret, "Thinking Like a Rat," *Angelaki* 20, no. 2 (2015): 121–34.

54. W. E. Castle, "The Domestication of the Rat," *Proceedings of the National Academy of Sciences of the United States of America* 33, no. 5 (1947): 109–17.

55. William H. McNeill, *Plagues and Peoples* (New York: Anchor Books, 2010), 45. The connection between rats and plague has been contested and rethought along multiple lines. Here the point is less to debate precise causal factors than it is to consider multispecies interaction as an assemblage. Victoria Gill, "Black Death 'Spread by Humans Not Rats,'" *BBC News,* January 15, 2018, https://www.bbc.com/news/science -environment-42690577.

56. Evans, "Blaming the Rat?"; Robert Sullivan, *Rats: Observations on the History and Habitat of the City's Most Unwanted Inhabitants* (New York: Bloomsbury, 2004), 1–7.

57. Sarah Ahmed, *The Cultural Politics of Emotion* (Edinburgh: Edinburgh University Press, 2014), 85.

58. Morton, *Dark Ecology,* 42.

59. Morton, 42.

60. Morton, 44.

61. Anthropocentric reason is like the search engine of agrologistics, seeking to determine the use-value of a thing in relation to these dictates.

62. There are important critiques of Morton's position. I do not address them in detail here because Morton's position is propositional for the development of this argument, but they are worth noting. See Stephanie Buongiorno, "Challenging the Dark Pools of Neoliberal Affect in Materialist Theories," *Graduate Theses, Dissertations, and Problem Reports* 5281 (2017): 24; Alexander García Düttmann, "Can There Be a Reconciliation with Nature?," *MLN* 133, no. 3 (2018): 709–19.

63. Douglas, *Purity and Danger,* 27.

64. Deleuze and Guattari, *A Thousand Plateaus,* 265–66.

65. Note here the contrast with dogs again, which arguably coevolved with human bands as hunters but then found new roles in agricultural domestication as protectors, herders, and so on.

66. Sullivan, *Rats.*

67. Ricci P. H. Yue, Harry F. Lee, and Connor Y. H. Wu, "Trade Routes and Plague Transmission in Pre-industrial Europe," *Scientific Reports* 7, no. 12973 (2017): 1–10.

68. Grant R. Singleton, Peter R. Brown, Jens Jacob, and Ken P. Aplin, "Unwanted and Unintended Effects of Culling: A Case for Ecologically-Based Rodent Management," *Integrative Zoology* 2, no. 4 (2007): 247–59.

69. Ahuja, *Bioinsecurities,* x, 195.

70. Graham C. L. Davey et al., "A Cross-Cultural Study of Animal Fears," *Behavior Research and Therapy* 36, no. 7–8 (1998): 735–50.

71. Rafi Youatt, "Interspecies Politics and the Global Rat: Ecology, Extermination, Experiment," *Review of International Studies,* May 2022, https://doi.org/10.1017/S0260210522000201.

72. Antoine Traisnel, *Capture: American Pursuits and the Making of a New Animal Condition* (Minneapolis: University of Minnesota Press, 2020), 2–15.

73. Youatt, "Interspecies Politics," 9.

74. Youatt, 12.

75. I follow Jack Halberstam here in understanding wildness as that which exists as the outside of an assumed order of things. Halberstam, *Wild Things: The Disorder of Desire* (Durham, N.C.: Duke University Press, 2020), 3.

76. Despret, "Thinking Like a Rat," 21–23.

77. Bruno Latour and Steve Woolgar, *Laboratory Life: The Construction of Scientific Facts* (Princeton, N.J.: Princeton University Press, 1986), 33–39.

78. This point is not intended to minimize that there is an exceptionally violent dimension to the control, killing, or destruction of rats as nonhuman animals. The situation described in this book is an exception to this broader rule, but an important one.

79. Michael Litzelman, "Benefit/Cost Analysis of U.S. Demining in Ethiopia and Eritrea," *Journal of Conventional Weapons Destruction* 6, no. 2 (2002): 50.

80. Deleuze and Guattari, *A Thousand Plateaus,* 324–89.

81. Weetjens, "How I Taught Rats."

82. "HeroRATs Save Lives," APOPO, 2020, https://www.apopo.org/en/herorats/herorats-save-lives.

83. "Animal Welfare."

84. Eva Short, "This Is How Bart Weetjens Taught Rats to Sniff Out Land-mines," *SiliconRepublic,* January 30, 2018, https://www.siliconrepublic.com/companies/bart-weetjens-inspirefest-2017-apopo.

85. Darcie DeAngelo, "Demilitarizing Disarmament with Mine Detection Rats," *Culture and Organization* 24, no. 4 (2018): 285–302.

86. DeAngelo, 287, 292–95.

87. DeAngelo, 299.

88. Weetjens, "How I Trained Rats," emphasis added.

89. In the language of Deleuze and Guattari, there are no individual heroes of humanitarianism but a series of virtual potentials that actualize and that sometimes form isomorphic ethical relations. In this sense, "humanitarian" is not a timeless ideal rooted in species or biological difference but a form of becoming in relation to others that could be actualized in the form of a human, dog, rat, or other nonhuman or posthuman. Deleuze and Guattari, *A Thousand Plateaus,* 425.

90. Jacques Rancière, *The Politics of Aesthetics,* trans. Gabriel Rockhill (New York: Bloomsbury, 2013), 89.

91. Her argument continues to demonstrate how humanitarianism structures risk so that failure, damage, and negative consequences of technology become legitimate outcomes of humanitarian deployments of technology. Katja Lindskov Jacobsen, *The Politics of Humanitarian Technology: Good Intentions, Unintended Consequences, and Insecurity* (New York: Routledge, 2015), 10.

92. Lefebvre, *Human Rights and the Care of the Self,* 3-4.

93. Connolly, *Facing the Planetary,* 12.

94. DeAngelo, "Demilitarizing Disarmament," 299.

95. Mark Duffield, *Posthumanitarianism: Governing Precarity in the Digital World* (Medford, N.J.: Polity Press, 2019).

96. "TB Detection Process," APOPO, 2020, https://www.apopo.org/en/what-we-do/detecting-tuberculosis/how-we-do-it/process.

97. "TB Detection Process."

98. Alan Poling, Bart Weetjens, Christophe Cox, Negussie Beyene, Amy Durgin, and Amanda Mahoney, "Tuberculosis Detection by Giant African Pouched Rats," *Behavior Analyst* 34, no. 1 (2011): 47–54; Georgies F. Mgode, Christophe L. Cox, Stephen Mwimanzi, and Christiaan Mulder, "Pediatric Tuberculosis Detection Using Trained African Giant Pouched Rats," *Pediatric Research* 84, no. 1 (2018): 99–103; Haylee Ellis, Christiaan Mulder, Emilio Valverde, Alan Poling, and Timothy Edwards, "Reproducibility of African Giant Pouched Rats Detecting

Mycobacterium Tuberculosis," *BMC Infectious Diseases* 17, no. 1 (2017): 298.

99. My thanks to Joyce Dinglasan-Panlilio for taking the time to discuss and explain VOCs. Lena Fiebig, Negussie Beyene, Robert Burny, Cynthia D. Fast, Christophe Cox, and Georgies F. Mgode, "From Pests to Tests: Training Rats to Diagnose Tuberculosis," *European Respiratory Journal* 55, no. 3 (2020): 2.

100. Fiebig et al., 2.

101. Fiebig et al., 2, emphasis added.

102. Fishel, *Microbial State,* 105.

103. DeLanda, *Intensive Science and Virtual Philosophy,* 90.

104. Despret, "The Body We Care For," 131.

105. Jacques Derrida, *Given Time I: Counterfeit Money,* trans. Peggy Kamuf (Chicago: University of Chicago Press, 2017), 13–30.

106. Jacques Derrida, *Memoires for Paul de Man* (New York: Columbia University Press, 1989), 149.

107. Derrida, 149.

108. Jacques Derrida, *Specters of Marx: The State of the Debt, the Work of Mourning, and the New International* (New York: Routledge, 2005), 18.

109. Puig de la Bellacasa, *Matters of Care,* 1–14.

110. Talal Asad, "Reflections on Violence, Law, and Humanitarianism," *Critical Inquiry* 41, no. 2 (2015): 397.

111. Jacques Derrida, "Force of Law: The 'Mystical Foundations of Authority,'" *Cardozo Law Review* 11, no. 5–6 (1990): 243.

112. Drucilla Cornell, "Rethinking Legal Ideals after Deconstruction," in *Law's Madness,* ed. Austin Sarat, Lawrence Douglas, and Martha Merille Umphrey, 147–69 (Ann Arbor: University of Michigan Press, 2006); Peter Goodrich, Florian Hoffman, Michel Rosenfeld, and Corenlia Vismann, eds., *Derrida and Legal Philosophy* (New York: Palgrave Macmillan, 2008); Jacques de Ville, *Jacques Derrida: Law as Absolute Hospitality* (New York: Routledge, 2011).

113. Marianne Constable, *Just Silences: The Limits and Possibilities of Modern Law* (Princeton, N.J.: Princeton University Press, 2005), 17.

114. Judith Butler, *The Force of Nonviolence: An Ethico-Political Bind* (New York: Verso, 2021), 27–66. For more specific discussion of the context of the laws of war, see Helen Kinsella, *The Image before the Weapon: A Critical History of the Distinction between Combatant and Civilan* (Ithaca, N.Y.: Cornell University Press, 2011).

115. Rajan Menon, *The Conceit of Humanitarian Intervention* (New York: Oxford University Press, 2016).

116. Asad, "Reflections on Violence."
117. Of course, the very notion of enforcement shows that this state of being wanting is, in fact, characteristic of all law. Jaywalking is often illegal, but it also wants for enforcement despite its existence in innumerable municipal codes.
118. Asad, 411.
119. Law's reliance on the metaphysics of presence also illustrates the ways in which agrologistics may constitute its conditions of possibility.

3. THE GIFT OF MILK AND THE CONTINGENCY OF HUNGER

1. Heifer International, "Baby Goat Explores Farm at Heifer International," YouTube video, 1:13, October 5, 2018, https://www.youtube.com/watch?v=gKDLk6xoaVo.
2. Orford, *Reading Humanitarian Intervention*; Oliver, *Carceral Humanitarianism*.
3. Donna Haraway explores a similar phenomenon in her analysis of crittercam videos. See Haraway, *When Species Meet,* 252.
4. James C. Scott, *Against the Grain: A Deep History of the Earliest States* (New Haven, Conn.: Yale University Press, 2017), 8; Sánchez-Villagra, *Process of Animal Domestication,* 43–49.
5. Maddie McKeever, "Honoring CPS and Heifer Project Work in Puerto Rico," Brethren Historical Library and Archives, August 15, 1944, https://www.brethren.org/bhla/hiddengems/honoring-cps-and-heifer-project-work-in-puerto-rico/.
6. McKeever.
7. For more on the "seagoing cowboys," see Holly Claycomb, "Bound for Adventure—Church of the Brethren, UN Relief Mission Share History of 'Seagoing Cowboys,'" *Altoona Mirror,* May 28, 2021, https://www.altoonamirror.com/life/area-life/2021/05/bound-for-adventure-church-of-the-brethren-un-relief-mission-share-history-of-the-seagoing-cowboys/.
8. Sheila J. Bryant, "Pay It Forward: The Heifer International Story," *Journal of Agricultural and Food Information,* October 11, 2008, 5–6; Peggy Reiff Miller, "First Seagoing Cowboy Takes Heifers to Puerto Rico," *The Seagoing Cowboys* (blog), August 20, 2014, https://seagoing-cowboysblog.wordpress.com/2014/08/20/first-seagoing-cowboy-takes-heifers-to-puerto-rico/.
9. My sincere thanks to Peggy Reiff Miller for guiding me through Heifer's

early history. Heifer International, *2021 Annual Report* (Little Rock, Ark.: Heifer International, 2021), 4.

10. Heifer International.

11. Ranganai Chidembo, "Influence of the Heifer International Pass on Programme on Livehoods of Households: The Case of Wanezi Ward in Mberengwa District of Zimbabwe," September 16, 2019, https://univendspace.univen.ac.za/handle/11602/1470; Heifer International, *2021 Annual Report*; Bryant, "Pay It Forward."

12. Linda Polman, *The Crisis Caravan: What's Wrong with Humanitarian Aid?* (New York: Picador, 2011).

13. Nicole Shukin, *Animal Capital: Rendering Life in Biopolitical Times* (Minneapolis: University of Minnesota Press, 2009), 8.

14. Robin Patric Clair and Lindsey B. Anderson, "Portrayals of the Poor on the Cusp of Capitalism: Promotional Materials in the Case of Heifer International," *Management Communication Quarterly* 27, no. 4 (2013): 537–67.

15. Heifer International, *The Future of Africa's Agriculture—An Assessment of the Role of Youth and Technology* (Little Rock, Ark.: Heifer International, 2021), 1.

16. "Food Security and Nutrition," Heifer International, https://www.heifer.org/our-work/work-areas/food-security-and-nutrition.html.

17. Clair and Anderson, "Portrayals of the Poor," 540.

18. James De Vries, "Passing on the Gift as an Approach to Sustainable Development Programmes," *Development in Practice* 22, no. 3 (2012): 373–84.

19. "12 Cornerstones," Heifer International, https://www.heifer.org/our-work/our-model/community-mobilization/cornerstones.html.

20. "12 Cornerstones."

21. Latour, *Reassembling the Social,* 30.

22. "12 Cornerstones."

23. "12 Cornerstones."

24. Clair and Anderson, "Portrayals of the Poor."

25. Haraway, *Companion Species Manifesto,* 4.

26. Laurent-Simpson, *Just Like Family,* 16–21.

27. Wolfe, *Before the Law,* 16.

28. Bourke, *What It Means to Be Human,* 90–112.

29. Regan, *Case for Animal Rights,* 196–99.

30. Elisabeth R. Anker, *Orgies of Feeling: Melodrama and the Politics of Freedom* (Durham, N.C.: Duke University Press, 2014).

31. Sánchez-Villagra, *Process of Animal Domestication,* 1–15.

32. Cynthianne Debono Spiteri, Rosalind E. Gillis, Mélanie Roffet-Salque, Laura Castells Navarro, Jean Guilaine, Claire Manen, Italo M. Muntoni et al., "Regional Asynchronicity in Dairy Production and Processing in Early Farming Communities of the Northern Mediterranean," *Proceedings of the National Academy of Sciences of the United States of America* 113, no. 48 (2016): 13594–99; Andrew Curry, "Archaeology: The Milk Revolution," *Nature* 500, no. 7460 (2013): 20–22; Richard P. Evershed, Sebastian Payne, Andrew G. Sherratt, Mark S. Copley, Jennifer Coolidge, Duska Urem-Kotsu, Kostas Kotsakis et al., "Earliest Date for Milk Use in the Near East and Southeastern Europe Linked to Cattle Herding," *Nature* 455, no. 7212 (2008): 528–31.

33. Sánchez-Villagra, *Process of Animal Domestication,* 90–122.

34. Kathryn Gillespie, "The Afterlives of the Lively Commodity: Life-Worlds, Death-Worlds, Rotting Worlds," *Environment and Planning A: Economy and Space* 53, no. 2 (2021): 280–95. See also Ryan Gunderson, "From Cattle to Capital: Exchange Value, Animal Commodification, and Barbarism," *Critical Sociology* 39, no. 2 (2013): 259–75.

35. Fahim Amir, *Being and Swine,* trans. Geoffrey C. Howes and Corvin Russell (Toronto, Ont.: Between the Lines, 2020); Hribal and St. Clair, *Fear of the Animal Planet.*

36. Sarat Colling, *Animal Resistance in the Global Capitalist Era* (East Lansing: Michigan State University Press, 2020), 35.

37. These dimensions are not actually separate but are distinguishable effects of cow–goat–human relations. Unpacking them thus helps to illuminate the different functional changes to cows and goats and their influence on human sociality.

38. O. T. Oftedal, "The Evolution of Milk Secretion and Its Ancient Origins," *Animal: An International Journal of Animal Bioscience* 6, no. 3 (2012): 355–68.

39. Etymology Online, s.v. "mammal," https://www.etymonline.com/word/mammal.

40. S. Blair Hedges and Sudhir Kumar, *The Timetree of Life* (Oxford: Oxford University Press, 2009), 459.

41. G. F. W. Haenlein, "About the Evolution of Goat and Sheep Milk Production," Small Ruminant Research 68, no. 1 (2007): 3–6.

42. Frost, *Biocultural Creatures,* 102–4; Raul Cabrera-Rubio, M. Carmen Collado, Kirsi Laitinen, Seppo Salminen, Erika Isolauri, and Alex Mira, "The Human Milk Microbiome Changes over Lactation and Is Shaped by Maternal Weight and Mode of Delivery," *American Journal of Clinical Nutrition* 96, no. 3 (2012): 544–51.

43. It is vital to note that this is not a "natural" or "species" trait but a socially specific practice, as the next section discusses in some detail. "Humans Were Drinking Milk before They Could Digest It," *Science,* January 27, 2021, https://www.science.org/content/article/humans -were-drinking-milk-they-could-digest-it; Michael Marshall, "Why Humans Have Evolved to Drink Milk," BBC, https://www.bbc.com/future/ article/20190218-when-did-humans-start-drinking-cows-milk.

44. "Food Security and Nutrition."

45. Hribal and St. Clair, *Fear of the Animal Planet,* 11.

46. Reviel Netz, *Barbed Wire: An Ecology of Modernity* (Middletown, Conn.: Wesleyan University Press, 2009); Alexander D. Barder, "Barbed Wire," in *Making Things International 2: Catalysts and Reactions,* ed. Mark B. Salter, 32–48 (Minneapolis: University of Minnesota Press, 2016).

47. Jackie Linden, "Lactation in Motion," The Pig Site, July 18, 2013, https:// www.thepigsite.com/articles/lactation-in-motion.

48. J. Barłowska, M. Szwajkowska, Z. Litwińczuk, and J. Król, "Nutritional Value and Technological Suitability of Milk from Various Animal Species Used for Dairy Production," *Comprehensive Reviews in Food Science and Food Safety* 10, no. 6 (2011): 291–302.

49. "Humans Were Drinking Milk."

50. Sánchez-Villagra, *Process of Animal Domestication,* 43–49; Scott, *Against the Grain,* 152.

51. While pursued for other reasons, the preference for homogenized milk is noteworthy as part of these aesthetics as well. Grove, *Savage Ecology,* 220.

52. Derrida, *Animal That Therefore I Am,* 28.

53. Clemens Driessen, "Animal Deliberation," in *Animal Politics and Political Animals,* ed. David Schlosberg and Marcel Wissenburg, 90–104 (London: Palgrave Macmillan, 2014).

54. Amy Hatkoff, *The Inner World of Farm Animals: Their Amazing Social, Emotional, and Intellectual Capacities* (New York: Stewart, Tabori, and Change, 2009); Jeffrey Moussaieff Masson, *The Pig Who Sang to the Moon: The Emotional World of Farm Animals* (New York: Ballantine, 2002).

55. M. Amills, J. Capote, and G. Tosser-Klopp, "Goat Domestication and Breeding: A Jigsaw of Historical, Biological and Molecular Data with Missing Pieces," *Animal Genetics* 48, no. 6 (2017): 631–44; Paolo Ajmone-Marsan, José Fernando Garcia, and Johannes A. Lenstra, "On the Origin of Cattle: How Aurochs Became Cattle and Colonized the World," *Evolutionary Anthropology: Issues, News, and Reviews* 19, no. 4 (2010): 148–57.

56. Oliver, *Technologies of Life and Death.*

57. Haraway, *When Species Meet*; Michel Serres, *The Parasite,* trans. Lawrence R. Schehr (Minneapolis: University of Minnesota Press, 2007).

58. Paula C. Pereira, "Milk Nutritional Composition and Its Role in Human Health," *Nutrition* 30, no. 6 (2014): 619–27.

59. As an example, see A. Elgersma, S. Tamminga, and G. Ellen, "Modifying Milk Composition through Forage," *Animal Feed Science and Technology* 131, no. 3 (2006): 207–25; K. Breuer, P. H. Hemsworth, J. L. Barnett, L. R. Matthews, and G. J. Coleman, "Behavioural Response to Humans and the Productivity of Commercial Dairy Cows," *Applied Animal Behaviour Science* 66, no. 4 (2000): 273–88.

60. Anthony John Weis, *The Ecological Hoofprint: The Global Burden of Industrial Livestock* (London: Zed Books, 2013).

61. Eric D. Schneider and Dorian Sagan, *Into the Cool: Energy Flow, Thermodynamics, and Life* (Chicago: University of Chicago Press, 2006), xiv, 6–11.

62. Peter Atkins, *Liquid Materialities: A History of Milk, Science and the Law* (London: Routledge, 2016).

63. Nissim Silanikove, Gabriel Leitner, and Uzi Merin, "The Interrelationships between Lactose Intolerance and the Modern Dairy Industry: Global Perspectives in Evolutional and Historical Backgrounds," *Nutrients* 7, no. 9 (2015): 7312–31.

64. Kendra Smith-Howard, *Pure and Modern Milk: An Environmental History since 1900* (New York: Oxford University Press, 2017); Atkins, *Liquid Materialities.*

65. Kim Severson, "A School Fight over Chocolate Milk," *New York Times,* August 24, 2010, https://www.nytimes.com/2010/08/25/dining/25Milk .html.

66. Susan Levine, *School Lunch Politics* (Princeton, N.J.: Princeton University Press, 2010).

67. Vasile Stănescu, "'White Power Milk': Milk, Dietary Racism, and the 'Alt-Right,'" *Animal Studies Journal* 7, no. 2 (2018): 103–28.

68. Spyros Spyrou, "Time to Decenter Childhood?," *Childhood* 24, no. 4 (2017): 433–37; Malkki, *Need to Help.*

69. Michel Foucault, *Security, Territory, Population: Lectures at the Collège de France, 1977–1978,* ed. Michel Senellart, trans. Graham Burchell (New York: Picador, 2007), 389.

70. Deleuze and Guattari, *A Thousand Plateaus,* 81.

71. Klaus von Grebmer, Jill Bernstein, Fraser Patterson, Miriam Wiemers,

Réiseal Ní Chéilleachair, Connell Foley, Seth Gitter, Kierstin Ekstrom, and Heidi Fritschel, *Global Hunger Index 2019: The Challenge of Hunger and Climate Change* (Dublin: Helvetas, 2019).

72. Georgios Paslakis, Gina Dimitropoulos, and Debra K. Katzman, "A Call to Action to Address COVID-19: Induced Global Food Insecurity to Prevent Hunger, Malnutrition, and Eating Pathology," *Nutrition Reviews* 79, no. 1 (2021): 114–16.

73. Mike Davis, *Late Victorian Holocausts: El Niño Famine and the Making of the Third World* (London: Verso, 2017).

74. Dieter Gerten, Vera Heck, Jonas Jägermeyr, Benjamin Leon Bodirsky, Ingo Fetzer, Mika Jalava, Matti Kummu, Wolfgang Lucht, Johan Rockström, Sibyll Schaphoff, and Hans Joachim Schellnhuber, "Feeding Ten Billion People Is Possible within Four Terrestrial Planetary Boundaries," *Nature Sustainability* 3, no. 3 (2020): 200–208; Charles C. Mann, "Can Planet Earth Feed 10 Billion People?," *The Atlantic,* January 23, 2018, https://www.theatlantic.com/magazine/archive/2018/03/charles-mann-can-planet-earth-feed-10-billion-people/550928/.

75. Harald Welzer, *Climate Wars: What People Will Kill for in the 21st Century* (Cambridge, Mass.: Polity Press, 2012).

76. In this sense, it functions similarly to human rights. Moyn, *Last Utopia,* 6.

77. Patricia Owens, *Economy of Force: Counterinsurgency and the Historical Rise of the Social* (Cambridge: Cambridge University Press, 2015); Alexander Barder, *Empire Within: International Hierarchy and Its Imperial Laboratories of Governance,* 1st ed. (New York: Routledge, 2015); Ann Laura Stoler, *Race and the Education of Desire: Foucault's History of Sexuality and the Colonial Order of Things* (Durham, N.C.: Duke University Press, 1995).

78. Renisa Mawani, "C Is for Cattle," in *Animalia: An Anti-imperial Bestiary for Our Times,* ed. Antoinette Burton and Renisa Mawani (Durham, N.C.: Duke University Press, 2020), 40.

79. Sonya Salamon, *Prairie Patrimony: Family, Farming, and Community in the Midwest* (Chapel Hill: University of North Carolina Press, 1995).

80. Graf, *Humanity of Universal Crime,* 47–58.

81. Marvin P. Miracle, "'Subsistence Agriculture': Analytical Problems and Alternative Concepts," *American Journal of Agricultural Economics* 50, no. 2 (1968): 292–310.

82. Robert L. Kelly, *The Lifeways of Hunter-Gatherers: The Foraging Spectrum* (Cambridge: Cambridge University Press, 2013); Scott, *Against the Grain,* 84.

83. Welzer, *Climate Wars*, 20–24.
84. Jenny Edkins, *Whose Hunger? Concepts of Famine, Practices of Aid* (Minneapolis: University of Minnesota Press, 2000).
85. Fishel, *Microbial State*.
86. Heather Davis, "Toxic Progeny: The Plastisphere and Other Queer Futures," *PhiloSOPHIA* 5, no. 2 (2015): 231–50.
87. Agamben, *Homo Sacer*.
88. E.g., Dominick LaCapra, *History and Its Limits: Human, Animal, Violence* (Ithaca, N.Y.: Cornell University Press, 2009), 165–67; Jared Sexton, "People-of-Color-Blindness: Notes on the Afterlife of Slavery," *Social Text* 28, no. 2 (103) (2010): 31–56; Matthew Calarco, "On the Borders of Language and Death: Agamben and the Question of the Animal," *Philosophy Today* 44, suppl. (2000): 91–97.
89. Martin Solich and Marcel Bradtmöller, "Socioeconomic Complexity and the Resilience of Hunter-Gatherer Societies," *Quaternary International* 446 (2017): 109–27; Mark Pluciennik, "The Meaning of 'Hunter-Gatherers' and Modes of Subsistence: A Comparative Historical Perspective," in *Hunter-Gatherers in History, Archaeology and Anthropology*, ed. Alan Barnard, 17–30 (New York: Routledge, 2004).
90. Bruno Latour, "On Technical Mediation—Philosophy, Sociology, Genealogy," *Common Knowledge* 3, no. 2 (1994): 29–64.
91. Gerten et al., "Feeding Ten Billion People"; Ronald Walter Greene, *Malthusian Worlds: U.S. Leadership and the Governing of the Population Crisis* (New York: Routledge, 1999), 59–65.
92. Deleuze and Guattari, *A Thousand Plateaus*, 18.
93. Quentin Meillassoux, *After Finitude: An Essay on the Necessity of Contingency*, trans. Ray Brassier (New York: Bloomsbury Academic, 2010); Eugene Thacker, *In the Dust of This Planet: Horror of Philosophy* (Washington, D.C.: Zero Books, 2011), 1:104.
94. Scott, *Against the Grain*, 8.
95. Friedrich Nietzsche, *On the Genealogy of Morals and Ecce Homo*, ed. Walter Kaufmann, Reissue ed. (New York: Vintage, 1989), 104.
96. Jason W. Moore, *Capitalism in the Web of Life: Ecology and the Accumulation of Capital* (New York: Verso, 2015).
97. Gilles Deleuze, *Difference and Repetition* (New York: Columbia University Press, 1994).
98. Jacques Derrida, "The Parergon," trans. Craig Owens, *October* 9 (1979): 3–41.
99. Jacques Derrida, "Force of Law: The 'Mystical Foundation of Authority,'" in *Deconstruction and the Possibility of Justice*, ed. Drucilla Cornell,

Michel Rosenfeld, and David Gray Carlson (New York: Routledge, 1992), 28.

100. Hribal and St. Clair, *Fear of the Animal Planet*; Pachirat, *Every Twelve Seconds*, 38–84.

4. HUMANITARIAN POLITICS ON A MULTISPECIES PLANET

1. Jack Halberstam, *The Queer Art of Failure* (Durham, N.C.: Duke University Press, 2011), 28, 47.
2. David Shannon, *Duck on a Bike* (New York: Blue Sky Press, 2002).
3. Bruno Latour, *We Have Never Been Modern*, trans. Catherine Porter (Cambridge, Mass.: Harvard University Press, 1993), 142–45.
4. Foucault, *Order of Things*, 278.
5. Don Kulick, "Human–Animal Communication," *Annual Review of Anthropology* 46 (2017): 357–78; Eileen A. Hebets, Andrew B. Barron, Christopher N. Balakrishnan, Mark E. Hauber, Paul H. Mason, and Kim L. Hoke, "A Systems Approach to Animal Communication," *Proceedings of the Royal Society B: Biological Sciences* 283, no. 1826 (2016): 20152889; William A. Searcy and Stephen Nowicki, *The Evolution of Animal Communication: Reliability and Deception in Signaling Systems* (Princeton, N.J.: Princeton University Press, 2010); Barbara Smuts, "Between Species: Science and Subjectivity," *Configurations* 14, no. 1 (2006): 115–26; Smuts, "Encounters with Animal Minds," *Journal of Consciousness Studies* 8, no. 5–6 (2001): 293–309.
6. Adams, *Sexual Politics of Meat*, 20; Derrida, *Animal That Therefore I Am*, 26; Kolbert, *Sixth Extinction*.
7. Isabelle Stengers, *Cosmopolitics* (Minneapolis: University of Minnesota Press, 2010); Latour, *We Have Never Been Modern*.
8. Soumya Iyenger, Pooja Parishar, and Alok Nath Mohapatra, "Avian Cognition and Consciousness—From the Perspective of Neuroscience and Behavior," in *Self, Culture and Consciousness: Interdisciplinary Convergences on Knowing and Being*, ed. Sangeetha Menon, Nithin Nagaraj, and V. V. Binoy, 23–50 (New York: Springer, 2017).
9. Slavoj Žižek, *The Plague of Fantasies*, 2nd ed. (New York: Verso, 2009), 3–44.
10. Jacques Derrida, *Points . . . : Interviews, 1974–1994* (Stanford, Calif.: Stanford University Press, 1995), 199.
11. Morton, *Humankind*, 143–45.
12. Deleuze and Guattari, *Anti-Oedipus*, 294.

13. Fishel, *Microbial State*; Eduardo Kohn, *How Forests Think: Toward an Anthropology beyond the Human* (Berkeley: University of California Press, 2013).

14. Jurgen Ruesch and Gregory Bateson, *Communication: The Social Matrix of Psychiatry* (New York: Routledge, 2017).

15. Robert W. Mitchell, "Bateson's Concept of 'Metacommunication' in Play," *New Ideas in Psychology* 9, no. 1 (1991): 73–87.

16. Bateson, *Steps to an Ecology of Mind*, 146.

17. Bateson, 276–83.

18. Morton, *Dark Ecology*, 13.

19. DeLanda, *A Thousand Years of Nonlinear History*, 185–211.

20. William Labov, "The Social Setting of Linguistic Change," in *Sociolinguistic Patterns*, ed. William Labov, 280–325 (Philadelphia: University of Pennsylvania Press, 1972).

21. Justin Eckstein, "Sound Arguments," *Argumentation and Advocacy* 53, no. 3 (2017): 163–80.

22. Morton, *Dark Ecology*, 14.

23. Gilles Deleuze, *The Logic of Sense*, ed. Constantin V. Boundas, trans. Mark Lester and Charles Stivale, reprint ed. (New York: Columbia University Press, 1990), 134–36.

24. Deborah Pugh, "Silence as a Form of Analytic Communication at the Level of the Basic Fault," *Psychodynamic Counselling* 3, no. 3 (1997): 279–89.

25. Bateson, *Steps to an Ecology of Mind*, 116–36.

26. Smuts, "Encounters with Animal Minds."

27. Schneider and Sagan, *Into the Cool*; Sagan, *Cosmic Apprentice*.

28. Sue Donaldson, "Animal Agora: Animal Citizens and the Democratic Challenge," *Social Theory and Practice* 46, no. 4 (2020): 709–35.

29. Colling, *Animal Resistance*, viii.

30. Donaldson and Kymlicka, *Zoopolis*.

31. Donaldson and Kymlicka, 12.

32. Eva Meijer, *When Animals Speak: Toward an Interspecies Democracy* (New York: New York University Press, 2019), 31.

33. Meijer, 17.

34. Massumi, *What Animals Teach Us*, 5–9.

35. Meijer, *When Animals Speak*, 57.

36. Meijer, 207.

37. Malkki, *Need to Help*, 110–15.

38. Samantha Power, *"A Problem from Hell": America and the Age of Genocide* (New York: Basic Books, 2002), 4–12.

39. Sophia Hoffmann, "Humanitarian Security in Jordan's Azraq Camp," *Security Dialogue* 48, no. 2 (2017): 97–112; Smirl, *Spaces of Aid*; Jocelyn Vaughn, "The Unlikely Securitizer: Humanitarian Organizations and the Securitization of Indistinctiveness," *Security Dialogue* 40, no. 3 (2009): 263–85.

40. Jane Bennett, *Influx and Efflux: Writing Up with Walt Whitman* (Durham, N.C.: Duke University Press, 2020), 4.

41. Weizman, *Least of All Possible Evils*, 1–17.

42. Bennett, *Influx and Efflux*, 8.

43. Bennett, 11, 50–53.

44. Bennett, 50, emphasis original.

45. Bennett, 65.

46. Thomas Nail, *Theory of the Earth* (Stanford, Calif.: Stanford University Press, 2021), 69–71.

47. On the dangers of animacy as a site of hierarchy, see Mel Chen, *Animacies: Biopolitics, Racial Mattering, and Queer Affect* (Durham, N.C.: Duke University Press, 2012), 2–5, 30.

48. Didier Fassin, "Humanitarianism as Politics of Life," *Public Culture* 19, no. 3 (2007): 499–520.

49. Adams, *Sexual Politics of Meat*, 28.

50. Wolfe, *Before the Law*, 45.

51. This problem is arguably endemic to all metaphysical approaches to life as a concept. Thacker, *After Life*.

52. This does not mean that humanitarianism does not appeal to biopolitical governance, just that it interlaces this with an image of the human apart from species identification.

53. Jacques Lacan, *Ecrits: The First Complete Edition in English*, trans. Bruce Fink (New York: W. W. Norton, 2006), 503.

54. This is not to say that these problems are not major concerns; rather, they occur as problems partly because of the way they are framed by humanitarian concepts.

55. Elisabeth Grosz, *Becoming Undone: Darwinian Reflections on Life, Politics, and Art* (Durham, N.C.: Duke University Press, 2011), 12.

56. If humanitarianism potentially contests an aspect of agrologistics in principle, it is that existence matters more than any quality of existing. However, in practice, humanitarianism often ends up merely protecting bare life. Morton, *Dark Ecology*; Scott, *Against the Grain*.

57. Mamdani, *Saviors and Survivors*.

58. Hannah Arendt's critique along these lines is perhaps the most widely cited. Arendt, *Origins of Totalitarianism* (New York: Harcourt, 1973), 269–75.

59. Félix Guattari, *The Three Ecologies,* trans. Ian Pindar and Paul Sutton, reprint ed. (New York: Continuum, 2008), 45.

60. Timothy Morton, *Ecology without Nature: Rethinking Environmental Aesthetics,* 1st ed. (Cambridge, Mass.: Harvard University Press, 2009).

61. Because no participant in humanitarianism is fully preoccupied with humanitarian concerns, there are certainly avenues for this to occur in humanitarian discourse, but humanitarianism remains recalcitrant.

62. Coetzee, *Lives of Animals,* 15–32.

63. "Global Animal Slaughter Statistics and Charts: 2020 Update," Faunalytics, July 29, 2020, https://faunalytics.org/global-animal-slaughter -statistics-and-charts-2020-update/.

64. Carys E. Bennett, Richard Thomas, Mark Williams, Jan Zalasiewicz, Matt Edgeworth, Holly Miller, Ben Coles, Alison Foster, Emily J. Burton, and Upenyu Marume, "The Broiler Chicken as a Signal of a Human Reconfigured Biosphere," *Royal Society Open Science* 5, no. 12 (2018): 180325; Gerardo Ceballos, Paul R. Ehrlich, Anthony D. Barnosky, Andrés García, Robert M. Pringle, and Todd M. Palmer, "Accelerated Modern Human–Induced Species Losses: Entering the Sixth Mass Extinction," *Science Advances* 1, no. 5 (2015): 1–5.

65. Derrida, *Animal That Therefore I Am,* 24.

66. Bourke, *What It Means to Be Human,* 73–93.

67. Timothy Morton, *Being Ecological* (Boston: MIT Press, 2018), 125.

68. Kathryn Gillespie, "An Unthinkable Politics for Multispecies Flourishing within and beyond Colonial-Capitalist Ruins," *Annals of the American Association of Geographers* 112, no. 4 (2022): 1108–22; Elan Abrell, *Saving Animals: Multispecies Ecologies of Rescue and Care* (Minneapolis: University of Minnesota Press, 2021).

69. Hebets et al., "A Systems Approach to Animal Communication"; Smuts, "Between Species."

70. Chris Cuomo, *Feminism and Ecological Communities: An Ethic of Flourishing* (New York: Routledge, 1998), 62; Donna J. Haraway, *Staying with the Trouble: Making Kin in the Chthulucene* (Durham, N.C.: Duke University Press, 2016); Gruen, *Entangled Empathy.*

71. Puig de la Bellacasa, *Matters of Care,* 47.

72. Thom van Dooreen, *Flight Ways: Life and Loss at the Edge of Extinction* (New York: Columbia University Press, 2014), 60, emphasis original.

73. Connolly, *Facing the Planetary,* 4.

Index

Page numbers in italics refer to illustrations.

Benjamin Meiches is associate professor of security studies and conflict resolution at the University of Washington–Tacoma. He is author of *The Politics of Annihilation: A Genealogy of Genocide* (Minnesota, 2019).